The Gift
of Good Land

Further Essays
Cultural and Agricultural

Wendell Berry

North Point Press
San Francisco 1981

Parts of "An Agricultural Journey in Peru," "The Native Grasses and What They Mean," "Horse-Drawn Tools and the Doctrine of Labor Saving," "Energy in Agriculture," "Solving for Pattern," "The Economics of Subsistence," "A Few Words for Motherhood," "A Rescued Farm," "Elmer Lapp's Farm," "A Talent for Necessity," and "New Roots for Agricultural Research" first appeared in *The New Farm* and are reprinted with permission of Rodale Press, Inc.

Parts of "An Agricultural Journey in Peru," January 1979; "The International Hill Land Symposium," April 1977; "A Good Scythe," January 1980; "Looking Ahead," July 1978; and "An Excellent Homestead," June 1979, are reprinted from *Organic Gardening,* Emmaus, PA 18049, with permission of Rodale Press, Inc.

A portion of "Sanitation and the Small Farm" was used in *Organic Gardening* in the article co-authored with Gene Logsdon, "Sanitation Laws Squeeze Out Small Producers." Reprinted from *Organic Gardening,* Emmaus, PA 18049, October 1977, with permission of Rodale Press, Inc.

"The Reactor and the Garden" first appeared in *Country Journal.*

"The Gift of Good Land" first appeared in *Sierra,* The Sierra Club Bulletin.

"Family Work" is reprinted from *Home Food Systems* © 1981 by Rodale Press, Inc. Permission granted by Rodale Press, Inc., Emmaus, PA 18049.

30622

for Gene Logsdon

Contents

Foreword

My previous book on agriculture, *The Unsettling of America*, sought to comprehend the causes and consequences of industrial agriculture within the bounds of a single argument: that agriculture is an integral part of the structure, both biological and cultural, that sustains human life, and that you cannot disturb one part of that structure without disturbing all of it; that, in short, though there may be specialized causes, there are no specialized effects. *The Gift of Good Land,* by contrast, is a collection of essays written separately for magazine publication. Though it is for that reason more random in form than the earlier book, it both confirms and somewhat improves upon it by presenting a greater number and diversity of exemplary practices.

In the course of the writing of both books, I have seen enough good farmers and good farms, and a sufficient variety of both, to convince me beyond doubt that an ecologically and culturally responsible agriculture is possible. Such an agriculture is now being practiced, productively and profitably, by a scattering of farmers all over the country. But there remain, I believe, two immediate obstacles to its success.

First, the discipline of farming has a low public standing.

A farmer is popularly perceived as a "hick," without the dignity, knowledge, or social respectability of a businessman or a member of a profession. The term "agribusinessman" is used partly as a euphemism for "farmer," and, not surprisingly, many farmers aspire to be "agribusinessmen." There is virtually no public appreciation of the complex disciplines necessary to good farming. Good farming is lumped in with bad farming as a low form of drudgery, not esteemed as the high accomplishment that it necessarily must be. This contempt is as readily found among the businessmen, journalists, academics, and experts who "serve" agriculture as anywhere else. Efforts to romanticize farmers only show the other face of the same contempt. It is the rule, I think, that we often romanticize what we have first despised.

The second obstacle is largely a result of the first. After a half century of industrial agriculture, farmers of any kind have become a tiny minority, and good farmers rare. To farm our land in the best way, to conserve it and keep it permanently productive, we need many more farmers than we have. Given the best of conditions, it would take a long time to get them. The best way to get farmers is to raise them on farms, but the seed stock has been drastically depleted. And for those who wish to come into farming from the outside, there are critical educational problems and few teachers. But conditions for going into farming now are discouraging if not prohibitive even for young people who know how to farm. At speculator's prices, land can hardly be made to pay for itself by farming. On top of that, the young farmer must pay usury to lenders, and buy equipment and supplies at costs rising much faster than the value of farm produce. It is a farmer-killing and a land-killing economy. Because the price of land is so disproportionate to the price of anything that can be grown on it, working farmers who own their land are worth more dead than alive. That is a joke now among farmers, and it is true.

Is there, in reality, such a possibility as "economy of scale" or "growth economy"? This question now presses heavily upon every enterprise of our livelihood. But upon agriculture, so near to the interests of culture and of life itself, it presses with the greatest weight. And it is from agriculture that we receive the most immediate answer: only if we are willing to sacrifice *everything* but money value, and count that sacrifice as no loss.

In agriculture, the economy of scale or growth directly destroys land, people, neighborhoods, and communities. (The same is true of industrial and urban "development" — though because the commotion is greater, the consequences may be less obvious.) And so good agriculture is virtually synonymous with small-scale agriculture — that is, with what is conventionally called "the small farm." The meaning of "small" will vary, of course, from place to place and from farmer to farmer. What I mean by it has much to do with *propriety* of size and scale. The small farm is defended in this book because smallness tends to be a prerequisite of diversity, and diversity, in turn, a prerequisite of thrift and care in the use of the world. In general, I believe, small farms tend to be diverse in economy, which is to say complex in structure; whereas the larger the farm, the more likely it is to specialize in one or two crops, to have no animals, and to depend on chemicals, purchased supplies, and credit. In agriculture, as in nature and culture, the more complex the system or structure (within the obvious biological and human limits), the more sound and durable it is likely to be. The present industrial system of agriculture is failing because it is too simple to provide even rudimentary methods of soil conservation, or to be capable of the restraints necessary for the survival of rural neighborhoods, and because it fosters a mentality too simple to notice these deficiencies.

There are, to be sure, urgent political and cultural arguments for the preservation of the small farm, and I hope that I have not overlooked or slighted them. But perhaps it is most necessary, now, to insist upon its *practical* justification: the overwhelming likelihood that its survival is indispensable to a sound, durable agriculture and a dependable food supply.

The practicality of the small farm may lie in the inherent human tendency to cherish what one has little of. I believe that land wasters always own or "control" more land than they can or will pay attention to. Some people, of course, will not cherish or pay attention to any land at all. But with land as with anything else, those who have a lot will tend to think that a little waste is affordable. When land is held in appropriately small parcels, on the other hand, a little waste tends to be noticed, regretted, and corrected, because it is felt that a little loss cannot be afforded. And that is the correct perception: it *cannot* be afforded.

Soil conservation, Henry Besuden wrote nearly forty years ago, "involves the heart of the man managing the land. If he loves his soil he will save it." There are fewer hearts involved now than there were then, and more soil is being lost. Wes Jackson, writing in 1981, is forced by consideration of the increased loss to the same perception; the cause of waste is "alienation from the land": "Where there is alienation, stewardship has no chance."

There is no want of evidence—some is given in this book; more is available—that the small farm, if properly ordered, equipped, and managed, is highly productive, kind to the land, and economically workable. This being so, we may ask why it has so few advocates in the colleges of agriculture, in government agricultural agencies, and in agricultural journalism.

The reason, I think, is a general one, and is to be found both in what we call our economy, and, because an economy is a cultural artifact, in our culture. For complex reasons, our culture allows "economy" to mean only "money economy." It equates success and even goodness with monetary profit because it lacks any other standard of measurement. I am no economist, but I venture to suggest that one of the laws of such an economy is that a farmer is worth more dead than alive.

A second law is that anything diseased is more profitable than anything that is healthy. What is wrong with us contributes more to the "gross national product" than what is right with us. Let us take a healthy marriage for example: a man and wife who produce from their own small farm or homestead or town lot as much as possible of what they eat, and provide on their own as far as possible for other needs; who therefore have work at home for their children; who therefore have "home life" and all that that implies. Such a couple may contribute immeasurably to the health of the nation, even to its solvency. But they are not good for the nation's business, for they consume too little.

If this man and wife were to get divorced, their contribution to the economy would increase spectacularly. Their household, with all its productive motives, means, and energies, would be dissolved, and its members would live by consumption. Their dependence on the industries of food, style, transportation, entertainment, and so on would be greater. So probably would their dependence on the industries of drugs, medicine, psychiatry, counseling, and the like. They would be worth far less to themselves, to each other, to their community, and to the world—but far more to the economy.

So it is in agriculture. The small Amish farms described

in this book, for example, are extraordinarily healthy. They are far more productive than consumptive, they support families and communities, and they preserve and improve the land. Surely our nation would be healthier if it contained several million such farms. And yet Amish farms have received virtually no attention from students of agriculture—the reason being, I think, that these farms are more profitable to the farmers than to those corporations whose livelihood has been the ruin of farmers. They consume small quantities of chemicals and commercial fertilizers; they use simple machines that last as long as machines ought to last; they use virtually no petroleum; they do not look upon indebtedness as a desirable temporary condition, much less a desirable permanent one; they grow their own sustenance.

But take, say, fifteen eighty-acre Amish farms and join them together in the ownership of an "agribusinessman," who will get rid of the livestock, take out the fences, buy the large machinery necessary to farm on a big scale, and plant all twelve hundred acres in corn or corn and beans. Health will decline in everything from the soil to the community; soil loss may rise as high as six bushels per bushel of corn. This farmer-as-"agribusinessman" will be a lifelong extravagant consumer of *everything* he needs, from fuel to fertilizer, from credit to extension courses in "stress management." He will be a good citizen of the economy. But whether he knows it or not, and sooner or later he will know it, this economy proposes to ruin him, as it has ruined millions of others, and sell him out to a bigger "agribusinessman" who wants to "handle" 2400 acres and help the economy even more.

Now let us say that a publishing company has some people in it who see the danger in this—the danger being the inevitable exhaustion of both soil and farmers, ultimately followed by starvation—and it undertakes to publish a magazine promoting agriculture of a better kind. It will discover that it has an abundance of important work to do; all the subjects and interests abandoned a generation ago by magazines like *Farm Journal* and *Successful Farming* lie before it. And it will discover that it has an audience waiting for it.

But sooner or later it will discover something else: from its own corporate financial point of view, it has the *wrong* audience. Its readers tend not to own large farms. They are mainly small farmers who do not have much money to spend, and who are apt to want articles on the purchase and repair of second-hand equipment, on the use of draft horses, on how to reduce production costs, on how to keep from buying things—not the sort to attract big advertisements for big tractors, without which the magazine is clearly not going to make big money. Maybe it could hang on for a few years, at a sacrifice, slowly building its readership, slowly collecting advertisers of small products, and finally make it. But that would be an act of faith and, no matter how well-intentioned, corporations do not take kindly to acts of faith, and they do not esteem sacrifice as a method. That is understandable.

The result, however, is regrettable: to prevent further sacrifice of money, there must be a sacrifice of readers. Those small farmers and would-be small farmers, so discouraging to advertisers, will have to be purged from the subscription list (by ignoring their needs), and the magazine will be aimed at larger, more consumptive farmers. And for this there is, in addition to the financial justification, a highly altruistic purpose: to convert the larger farmers who, after all, are abusing far more land than the small farmers.

It is, as I say, understandable but regrettable. The small farmers lose yet another defender. And if conversion is the aim, the small farmers are easier to convert, and have more to gain by conversion, than the larger ones. The larger farmers, after all, are usually the most dependent on the farming system from which it is desirable to convert them. But there is a related sacrifice that is perhaps even more damaging. To think well about farming, you need to be familiar with the whole spectrum of its current means and methods. To disregard the small farmers is to forego one of the essential means of understanding the large ones. To disregard the comparatively independent farmers, such as the Amish, is to ignore the only available standard by which to measure the performance of the critically dependent ones. It is to turn aside from the critical questions of size and scale. It is, in short, to fall into the same trap that the conventional farm publications have already fallen into. The farmer who grosses $100,000 a year becomes more attractive than the farmer who *nets* $20,000 a year. That the $20,000 may be netted on a farm much smaller, much better farmed and cared for, is simply left out of account.

The same flawed solution to the same dilemma seems to hold, less understandably, in the colleges of agriculture: the same dependent's allegiance to "economy of scale" and "purchased inputs." Anyone who listens to an early morning farm program on the radio will be aware how seldom a university expert will propose a remedy that is not for sale. What they have accomplished is the virtual substitution of credit for brains.

Some months ago I attended a clinic for farmers on the growing of alfalfa. The university experts—a different expert for each problem of alfalfa—held forth in the front of the room. In the back of the room were the displays and salesmen of seed and chemical companies. The farmers, significantly, were in the middle, expected to run directly from advice to purchase. The farmers' own tax dollars were thus being spent to advise them to patronize corporations that would benefit far more than the farmers from that patronage.

Because of this virtually self-enforcing orthodoxy—this plutocracy masquerading as "agribusiness"—the defenders of the small farm are in decline along with the small farmers. It is not a well-paying defense. To advocate the survival of the small farm is necessarily to advocate a set of solutions that cannot be sold or bought, but are the free birthright of a community—and there are few employers for that. Nevertheless, the small farms still have defenders—and, I trust, will continue to have them.

To these defenders I want to suggest that it may be impossible to defend the small farm by itself or for its own sake. The small farm cannot be "developed" like a product or a program. Like a household, it is a human organism, and has its origin in both nature and culture. Its justification is not only agricultural, but is a part of an ancient pattern of values, ideas, aspirations, attitudes, faiths, knowledges, and skills that propose and support the sound establishment of a people on the land. To defend the small farm is to defend a large part, and the best part, of our cultural inheritance.

Defenders of the small farm (to use only the most immediate example) must take care never to use the word "economy" to mean only "money economy." We must use it to mean also—as the origin of the word instructs—the order of households. And we must therefore judge economic health by the health of households, both human and natural.

It would be misleading for me to take more than partial credit for the essays in this book, which could not have been written without the good work and the generosity of a number of people who consented to be my subjects, hosts, guides, and teachers. I am happy to acknowledge my indebtedness to Steve and Peggy Brush, Karen Reichhardt and Gary Nabhan, Francisco Valenzuela, Fred and Alice Kabotie, Tim Taylor, Bill Martin, Earl Spencer and his family; to Nick Coleman, Wally Aiken, Tom and Ginny Marsh, Elmer Lapp, Henry Besuden, Wes and Dana Jackson, Perry Bontrager, Ervin Mast; to Bill, Richard, Orla, Mel, and Wilbur Yoder—and, as before, to Maury and Jeanine Telleen.

To Steve Smyser, editor of *The New Farm* during my time on the staff, I am grateful for innumerable kindnesses, and for good advice, criticism, and companionship.

My gratitude to Gene Logsdon is acknowledged in the dedication. I owe him the sort of debt that can be neither contracted nor repaid except in friendship. There is hardly a notion in this book that did not first pass in conversation between us. Our agreement has been broad enough that our differences have been invariably instructive, and usually entertaining.

During my absences from home, which this work required, my wife and son kept the farm, doing my work in addition to their own. Whatever I could say to them in thanks would be insufficient. My wife also typed these essays in several drafts, serving in the process as their most alert and understanding critic.

Whatever is good in this book is owed in large measure to the work and the goodness of all those people. Of its faults, I am sole author.

I

1

An Agricultural Journey in Peru
(1979)

I have travelled quite a bit in the last several years and ought to be getting used to it, but I never find it easy to leave home. Leaving for Peru in the middle of June was especially difficult. For anyone with a farm, a two-week absence at that time of year will be problematical. For a raspberry lover to leave his raspberry patch at that time of year borders on calamity. I took a last look at the green berries, wished them a slow ripening, and headed for the airport.

Hard as it was to leave, I had at least three good reasons for wanting to go. (1) I wanted to see an *ancient* American agriculture, and in the Peruvian highlands a native agriculture has been carried on continuously—with modifications, of course—for something like 4500 years. (2) Principally from the work of Stephen Brush, anthropologist at the College of William and Mary, I knew that this agriculture involved great skill and ecological sensitivity in the use of steep, rocky, and otherwise "marginal" land. I wanted to see firsthand the practices that had enabled farming to survive so long in such a place. (3) I wanted to understand the pressures imposed on the traditional agriculture, first by the Spanish Conquest and colonization beginning in 1532, and now by population growth, migration to the cities, industry, commerce, and various other aspects of "modernization."

All this depended on Steve Brush's willingness to act as my host, guide, teacher, and interpreter. I had a general notion of what to look for, but I did not know where to look; I did not know the country or the languages. And so at the end of a long Saturday's travel, when I had made my way past the various officials and inspectors at the Lima airport, it was a relief to see Steve waiting for me.

It was dark by then. We carried my bags to the car, where I met Steve's wife Peggy and their four-year-old son, Jason, who was asleep in the back seat.

Just after leaving the airport we drove past a large settlement of squatters, homeless people from the highland villages, who have built a sizable suburb of straw or reed mats on a vacant plot intended for a public school. The huts, joined like row houses, are without sewers, plumbing, or electricity. Their major advantage is that Lima's climate, though often cloudy and damp, is virtually rainless and never cold. These improvised dwellings can be roofed with the same mats that they are walled with. In one of them, as we passed, I saw an open fire flickering. It would, I thought, be an uneasy place to sleep.

After an excellent meal at the house of Steve's research assistant, Alejandro Camino, and his wife Patel, the Brushes drove me to my *pensión,* Residencial Miraflores, where I was to stay until Monday, when Steve and I would go to the city of Huancayo in the highlands.

The next morning the Brushes came by at eleven o'clock to drive me around Lima. This was an election day and the traffic was heavy. People do not vote in their neighborhoods or precincts, but are assigned voting places which may be a long way from their homes. And so they have to travel to vote. A high percentage of the vehicles are badly run down and patched up. Worn or poorly tuned engines make a great stink and haze. And Peruvian traffic, to an outsider anyhow, is a bewilderment, making its way less according to rules than by a mixture of aggression, bravado, and desperate improvisation. At one point I saw a large bus easing cautiously into a crowded intersection, the traffic swirling, honking, and fuming all around it; it was going backwards! There are policemen in booths at the busier intersections, but they tend to look futile and forlorn, as if in need of rescue.

The people do not brave the election traffic necessarily because of any political loyalty or enthusiasm. They vote because they *must* do so or lose the perquisites of citizenship. All over town they were waiting outside of the polling places in long lines, under the vigilance of soldiers with guns who sat or stood in armored vehicles.

But interesting as were these various manifestations of civic order, they were not the main purpose of our excursion. Steve and Peggy, with what I would later see to have been admirable calculation, were taking me to see the slums, the *barriadas,* of Lima. They wanted me to realize that what I would see in the highlands should not—indeed, could not—be understood apart from the crowdedness, the poverty, the needs, and the ambitions of this troubled, troubling city.

The population of Lima is now about 4,500,000. This figure has more than doubled in the last ten years. Two of the subdivisions of the city, Miraflores and San Isidro, are attractive and well kept. These are the upper-class sections, and may be said to stand for the "modern" or "American" ambition of the country. To a visitor from the United States, this ambition is obvious and familiar; it is the mixture of comfort, convenience, affluence, and petroleum power that we call "the American way of life," one of the necessities of which is to ignore or forget "how the other half lives."

In Lima the other half—or the other 80 percent—lives the best it can, which is to say poorly. The tides of people flow in, bearing down any possibility or pretense of "orderly growth." The *barriadas* are thrown together without plan, according merely to the principle of accumulation, sprawling out over the dead, pinkish-gray desert around the city. Often there are no public services of any kind—no water, no sewers, no electricity, no garbage collection. The houses, scattered over the slopes of otherwise lifeless hills, are small, built of adobe or brick or straw mats. In places little terraces have been leveled, dug out of the hillsides, sometimes walled with rock, to afford building sites. In contrast with the affluent neighborhoods, which seem almost deserted, the slums are crowded. People are everywhere, standing around, hurrying this way and that, visiting in groups, waiting for buses, working on old cars. There are clearly many too many children. Here and there a few people and many bony dogs sort through the garbage, which is scattered and dumped in piles everywhere along the roadsides.

But most of these people looked well enough fed and clothed. They looked clean—a touching accomplishment, considering that water must be almost as limited a commodity for them as gasoline. And over and over again we saw small but unmistakable signs of self-respect: ornamental designs on the faces of houses, small dooryards with trees or shrubs or blooming flowers—watered at what cost or sacrifice?

"What brings them here?" I kept asking myself. What can induce people who so recently belonged to the open countryside to come and live in such places? Undoubtedly there are negative causes, the pressure of population growth in the villages being the greatest. But there must be attractions too. For instance, one observes here as in the United States the by now classical association of poverty,

TV, and an old car. The lowliest hut in a Lima slum sends heavenward its TV antenna, and has a battered Chevrolet parked at the door. The TV and the car, I think, stand for the popularized industrial aspiration, the desire for consumer goods, diversion, mobility. The poverty stands simply for the inevitable failure of supply: Everybody in the world cannot have more than enough of everything. It is a self-defeating aspiration, for it means surfeit for some and frustration for many. If it is failing in the industrial cities of the United States, I thought, how can it hope to succeed in Lima?

There is no welfare or unemployment insurance. The government's social security program pays no benefits; its entire income goes to pay for its large new office building. In consequence, Steve says, the slum dwellers display great inventiveness in getting a living. For instance, we passed a place where several oblong, shallow holes were dug into the roadside. This was a muffler shop. A driver straddles his car across one of the holes; the mechanic crawls under to work on the muffler.

After our drive, I sat down with Steve in his apartment to hear about his year's work in the Andean potato fields. Urban people, he said, generally assume that the peasant farmers are ignorant, that they have no systematic knowledge or method, that they don't understand the morphology of plants, that they name varieties whimsically or arbitrarily. He believes, on the contrary, that they have a great deal of sophistication in their choice and use of varieties and in their cropping systems. The work that he had just completed tended to confirm his earlier belief, resting on several previous years of work in Peru, that these people farm with a highly refined ecological sensitivity, competently attentive both to the capacities and limits of their fields and to the relationship between crop variety and place. This respect for the methods and the intelligence of the peasant farmers is the foundation of his work, which thus opposes the destructive stereotype of the farmer-as-ignoramus. For almost a year he had been studying the use and cultivation of native potato varieties and local taxonomy, even growing some of the native varieties himself. He had spent many hours identifying and counting the varieties in various fields. As we talked, we had his complexly coded maps and charts around us on the floor.

Variety is the security of agriculture, as of biology. Unlike the "scientific" agriculturists who give priority to "efficiency," the Andean farmers' first principle is variety. It is the ancient wisdom of putting the eggs into several baskets; in a season or a field in which one variety perishes, another, or several others, may thrive. And so the focus—and the difficulty—of Steve's work has been diversity. He recalled one field, about the size of an ordinary living room, in which he found forty-six different potato varieties.*

Over the centuries, he said, the peasant farmers have done a lot of selecting of varieties. And the varieties thus developed suit their needs well. They will grow the "improved" potatoes developed by plant geneticists, but 90 percent of these are grown for sale. The farmers don't keep and use them because they don't like their taste—they are too watery—and because they don't keep well. The introduced varieties yield possibly 30 percent more than the native, but they may be 90 percent water, whereas the native varieties will be 80 percent.

At this point I was reminded of a paragraph in a travel

---

*Such mixed fields are grown solely for household use, the diversity of varieties having an obvious attraction for people who eat a lot of potatoes. And not all plants, of course, will successfully grow so many varieties. The point here is that this farmer had a repertory of at least so many varieties.

guide that I had read before leaving home. I quote it here as an example of agricultural thinking unfortunately not limited to the authors of travel guides:

> You'll probably notice the favorite Indian food, potatoes, although it will take a little effort to recognize the tiny, apricot-sized tubers for what they are. In this high altitude and hostile Andean climate this is about as big as the potatoes ever grow. Approximately a half dozen would be required to equal one typical American potato. Incidentally, the Peace Corps boys have been trying assiduously to encourage the Indians to try another variety of potato, which grows better in a cool climate and at high altitudes, but the Indians have shown little inclination to go along with innovations. They prefer their old, familiar variety of potato, however bad it may be.

That is wrong both in particular and in general. In particular, it assumes that there is only one native potato variety in a region where varieties are more numerous than anywhere else (estimates range from "well over 400" to more than 2,000). In general, it assumes, conventionally but too readily, that bigger is better. A small mathematical operation is necessary here: if an "improved" variety yields 30 percent more than a native variety, but has only half as much dry weight (10 percent as opposed to 20 percent), how much improvement do you have? And bigness, as Daniel Gade points out in *Plants, Man, and the Land in the Vilcanota Valley of Peru,* may offer no advantages: "the peasants generally value numbers above size, since at high elevations less time and heat are needed to thoroughly cook a small seed or tuber than a large one."*

The several reasons for rejecting the new varieties, then,

---

*Steve Brush has recently informed me that the new commercial varieties also "degenerate"—probably because of a virus—in the sense that yield and potato size are reduced in successive plantings, so that new seed potatoes must be purchased every one to three years—a heavy expense for the farmers.

have mainly to do with *quality*. It is significant that these people are farmers in the most ancient sense of the word: that is, they are first of all subsistence farmers; they eat what they grow, selling only the surplus. Where the "improved" or high-yield varieties have caught on and replaced the native varieties is on the larger farms that produce potatoes only for sale. Again, some elementary calculation is in order. If you sell all you grow, you want to sell as much as possible; your interest, then, is in *quantity*. If, on the other hand, you intend to eat at least a part of what you grow, you naturally want it to be as *good* as possible; you are interested, first of all, in quality; quantity, important as it is, is of secondary importance.

Might it not be, I thought, that subsistence farming is the very definition of *good* farming—not at all the anachronism that the "agribusinessmen" and "agriscientists" would have us believe? Might it not be that eating and farming are inseparable concepts that belong together on the farm, not two distinct economic activities as we have now made them in the United States? Is not "agribusiness" the name of farming divorced from eating? (I thought of my raspberry patch, where I have done not a minute of work that was not inspired by the thought of eating; where work makes hunger, and hunger justifies work; where quality is never so much a principle as a pleasure.)

Allen Benjamin, who works as an economist for the National Corn Program at the Agrarian University in Lima, later joined our conversation. He said that there is now a bona fide effort on the part of potato scientists to understand the traditional farming—its virtues as well as its shortcomings. I was also interested to hear that Allen's group is conducting on-farm trials of new varieties, the work being done by the farmers themselves. This seems far better than trying them out in optimum conditions on "experimental farms," and leaving the farmers to discover the drawbacks on their own. In breeding new varieties of corn, Allen said, insect, disease, and drought resistances have been given less attention than yields. And so the new strains, though purer than the old, are "not that much better."

I asked Allen why the protein deficit in the Peruvian diet is not supplied with beans—and I received a lesson in the delicacy of Andean agriculture in the face of a growing demand for food. Disease, he said, is the problem. To increase food production, the planting dates of corn have been advanced, so that the growing seasons at different altitudes overlap, thus providing year-round sustenance to the aphids, which are the carrier of a bean virus. (According to Daniel Gade, there are two other considerations that limit the use of kidney beans in the highlands: the cool nights retard their growth, and when they are dried it takes too long to cook them in an area where fuel is scarce.)

The advanced planting dates also disrupt the relation of crops to rains, hence to fungi. Because of diseases, the yields of both corn and beans are lower now than they were fifty years ago.

Thus, as often in dealing with such problems, we are hard up against a paradox: there are more and more people, needing more and more food, but the measures intended to increase food production also increase pests and diseases, which *decrease* food production.

Allen said that Peru imports 600,000 tons of North American corn for chicken and hog feed. It also imports most of its wheat. But "thousands of tons" of Urubamba Blanco (a variety of corn) are exported annually, mostly to Japan.

The United States subsidizes the transport of wheat to Peru in order to get rid of its surplus. This drives down the price of Peruvian wheat. Peruvian wheat growers, as a result, now grow barley, which they sell to brewers for the making of beer. This is the result of Public Law 480, Food for Peace, which provides "cheap food for hungry countries," but can seriously disrupt local production. Which is paradox number two: charity to a hungry country makes it dependent on foreign imports for bread, but makes it independent in the production of beer.

Steve said that probably 80 percent of the potatoes that Lima eats are "white" (high-yield) potatoes. Perhaps one-third to one-half of these are produced on the coast by big producers on plots of ten acres or more. The highland farmer, by contrast, cultivates plots of perhaps half an acre. The attention of the agricultural programs and researchers drifts toward the larger producers in the interest of "efficiency" and cheapness. Steve's attention drifts to the small producers. This is hardly a frivolous or a sentimental preference, as our forthcoming travels in the highlands would show.

Steve identifies four general categories of potatoes: table potatoes; *chuño* (bitter) potatoes, used only for freeze-drying; weed potatoes, which are "semi-domestic," usually growing in association with corn; and wild ("fox" or "frog") potatoes. That cross-breeding can occur among, as well as within, these categories explains the rich genetic diversity of potatoes in the Andes. One of the principles of this agriculture is that something new and good may turn up, in the course of nature, at any time. Andean farmers have been alert to that possibility for more than four thousand years.

On Monday we were to drive up to Huancayo, in the Mantaro Valley, which would be our headquarters during several days of agricultural travels. But in what I would learn to recognize as Peruvian fashion, we spent the entire morning on errands, getting ready. We had to make reservations for the remainder of my trip, get some money changed, and rent a car. The reservations didn't take much longer than I expected. The other two operations entailed surprising complexities. To rent a car we had to go to two different offices (of the same agency), sign many documents, and leave my passport as hostage.

In our white Volkswagen, thus rented and ransomed, we committed ourselves again to the fates of Lima traffic. Steve remembered this story: a driver in Lima stopped for a red light, and was promptly rammed from behind by another car. A policeman arrived, studied the situation, and pronounced the first driver in the wrong: he should not have stopped; there had been no traffic on the cross street.

With the red tape, the fog, the slums, and the traffic of Lima behind us, we felt positively liberated when at last we emerged into the sunlight of the Rimac Valley and saw the mountains before us. The road follows the Rimac River up through its valley, crosses the 15,982-foot Ticlio Pass, and then descends along the Mantaro River to Huancayo at about 11,000 feet. First there are the commercial farms on the broad flood plain of the lower Rimac; then, as the valley grows narrower and steeper, the fields become smaller, tucked in among boulders, or walled against the currents of the river. And already the dominant theme of the mountain agriculture is established: great care and frugality in the use of land; every available scrap of land, no matter how small, is painstakingly used and conserved. After the lifeless mountains near Lima, these neat little plots of corn, alfalfa, and fruit trees are strangely touching. For a while all the farming is by irrigation, the band of green beginning abruptly where the ditches pass along the slopes. And then, as we climb into better watered country, the characteristic small fields of potatoes or grain begin to appear on the mountainsides.

The confluences of the traditional agriculture with

American "agribusiness" are sometimes grotesque. Steve said that Aldrin, Dieldrin, and DDT, all now banned in the United States, are sold freely in Peru by their U.S. manufacturers. And here they are used inexpertly and incautiously. They are applied by hand from gunny sacks shaken over the rows. Some farmers, for economic reasons, used 30 percent less than specified—and so immunized the pests, and killed off the pests' parasites.

Above the crop zone is the zone of high pasture, where cattle, sheep, llamas, and alpaca graze the tufts of ichu grass.

We crossed, light-headed, over Ticlio Pass, at almost 16,000 feet still not up to snow line, and started our descent. Now we were in mining country. Since Pizarro arrived in 1532, looking for silver and gold, Peru has borne the serfdom of the mines. Wherever I have seen it, no matter who owns it, mining country is a colony. Whether they are "domestic" or foreign, the interest of mine owners is in what is under the ground; they respect no living thing that is on top of it. The mining towns I saw in Peru were very similar, I thought, to the coal camps of Eastern Kentucky. They had the same minimal housing, the same woeful drabness, the same pragmatic ugliness, the same uprootedness of people, the same alienation or absenteeism of authority.

The town of Oroya is the site of a huge zinc and lead smelter. Its sulfuric-acid fumes have killed all vegetation in the valley downwind for perhaps five miles; the terraces of old potato fields along the slopes are as lifeless as tombs. And below this smelter the Mantaro River is worse polluted than any I have seen, its rust-colored water now useless for irrigation, for which it was used before the industrialists came. Peru, in desperate want of food, is said to need its industrial income to buy food from other countries—but its industry reduces its ability to grow its own food. Paradox number three.

We escaped the industrial desert into healthy country. It grew dark and a full moon rose, hanging exactly over the notch of the valley ahead of us.

Huancayo, a city of about 150,000 people, is not particularly comely. Steve said it is like a frontier town and, though it is not on a frontier, it does have about it what I thought was a sort of frontier feeling. It seems raw and incomplete. There are unfinished buildings everywhere. The streets, at any time I saw them, were crowded, as if it were always Saturday night. People parked their cars not only along the sides of the main street, but also down the middle, between the opposing lines of traffic.

The day after our arrival, Tuesday, we visited the country around the villages of Aymara and Pasos where, Steve told me, he had identified seventy-seven different potato varieties.

We drove out southeast (downriver) from Huancayo, on a road that went dead straight along the valley. The road was lined with houses and the mud-walled enclosures of farms. There was a lot of livestock along the road: pigs, cattle, horses, burros—some staked out to graze, some being herded or driven along the side of the road. After a while we climbed up past a village on a steep, winding gravel road. The farmers were harvesting wheat, barley, oats, broad beans, field peas.

We stopped and talked with a man who was cutting a little field of bearded wheat with a sickle. He was chewing coca, working alone. He treated us hospitably—showed us his sickle and talked about his crop and his work. The sickle blade, serrated much like a mowing machine section, is sharpened with a file. The cut grain is gathered by handfuls and laid in sheaves on the ground, the heads in alternate handfuls turned first one way and then another. Later the sheaves are bound with straw, loaded on burros or horses, and carried to the threshing floor. Depending on the crop, the grain is either tramped out by oxen or burros or beaten out with flails. The grain is tossed into the wind to winnow it and then bagged for transport by pack animals. I did not see anywhere a wagon or a cart.

As we talked to the reaper, we could see threshers starting work at several farmsteads below us. In a nearby field another reaper was at work, also alone, whistling to himself. And farther down were the tile roofs and eucalyptus trees of the village. Above us, along the road, people were climbing to their fields, taking their stock along to forage near the croplands while the work went on. In one family that passed with its animals, a little boy was riding a black pig.

The fields were small, bordered with stones and brushy plants that help to protect them from both grazing animals and erosion. The edges of some of them had been planted with a species of tall-growing bitter lupine that stock will not eat, which also helps to protect the crops.

The pasture is nowhere very good—judged, say, by the standards of the Ohio Valley—and yet there seemed a fairly heavy livestock population. Animals, that morning, were everywhere, grazing the fallowed fields, gleaning the harvested ones. And they were remarkably well controlled. Some were staked out, on ropes made, as nearly as I could tell, of strips cut from the sidewalls of tires. I saw, to my surprise, a lot of hogs staked out in harnesses like those used for dogs or with ropes tied around their necks. That testifies rather eloquently to the condition of the hogs; a *fat* hog, for all practical purposes, has no neck. But most of the stock ran loose, under the vigilance of herders, usually women or children, who controlled them by hissing and hollering, or throwing rocks or clods, or hurling rocks with their long slings of braided llama wool. No apparent effort is made to keep the kinds of animals apart. I saw various mixtures of horses, burros, hogs, sheep, llamas, alpacas.

None of the species of stock that I am familiar with seemed to have been bred selectively, one obvious reason being the failure to castrate inferior males. I never saw so many full-grown bulls. This was my first acquaintance with llamas and alpacas, but the quality of these native species looked far better than that of the hogs, sheep, and cattle.

These hillsides, like nearly all that we saw, were very steep, and much more intensively farmed than I had expected. There is some erosion, bad in certain places where the water has apparently been diverted around fields, but in general the land seems well conserved. It is necessary to keep in mind that these fields have been farmed and have lasted a long time—some from before the Incas. The ground is worked and planted in varying patterns according to the availability of water and suitable ground. The rows are laid off mostly up and down the slope, which at first seems contrary to sense—and then makes more sense the more you study the problem. On such steep slopes, considering that the rows are "hilled up" to a height of perhaps as much as ten inches, contoured rows would almost certainly erode worse under heavy rains than these downhill ones. The water would collect behind the mounded, loosened dirt of the contoured rows and break through, carrying the dirt with it. The downhill rows, on the other hand, let the water out quickly, not allowing it to accumulate anywhere, and it runs over the packed, unworked earth between the rows.

But in some fields, not many, the rows went across the slope; in others they went downward on a slant. Some of these slanted rows zigzagged across the slope. The purpose of this slanting, I guessed, is to slow down the runoff to

allow the ground to absorb more water. The fault of this pattern was that where the Vs pointed downward there would be a wash. A much more satisfactory pattern, affording both water retention and erosion control, was a zigzag *down* the slope.

Many of these highland fields are still broken with the foot plow (in Quechua: *chakitaklla*). Weeding occurs only after the plants are well established (six to eight inches high); the weeds are thus left undisturbed to hold the soil until the roots of the crop plant can take over the job. In some fields the potatoes are planted directly into the sod; a planting hole is opened with a foot plow, and the seed potato and a little manure are dropped in and covered, the ground not being worked until about six weeks later. In any talk of soil conservation in the Andes, it is necessary to consider the quality of Andean sod, which is extremely tough and fibrous, much harder to shake apart than the sod I am familiar with in Kentucky. In cultivating, the chunks of sod seem often to be merely inverted and left more or less intact during the growing season.

Above the village and its plantings of eucalyptus, the slopes were bare of trees. This is characteristic of the Andes, and to one used to the forested slopes of North American mountains it is the strangest thing about them. Civilization is old in the Andes, and so apparently are ecological problems. The following observation from Carl O. Sauer's *Agricultural Origins and Dispersals* should be borne in mind as an antidote to the tendency to idealize the native culture, careful in many ways as it has been:

> The highlands are miserably poor in fish, game, and fuel. Gilmore suggests extermination by hunting to explain the occurrence of several extinct animals found in archeological sites near Cuzco. Guanaco and Vicuna have disappeared from the larger part of their earlier range. These items indicate that man got seriously out of ecologic balance in highland Peru and Bolivia, that he overhunted animals as he overcut wood for fuel.

The highlands once may not have been as bleak as they are now. The steady expansion of the Inca state may have been in part due to the need of more and better food for the protein- and fat-starved central highland.

We had been climbing the side of a high ridge which crested at about 14,000 feet. At the top we drove to a microwave station on a sort of knoll above the ridgeline, and climbed to the top of a jumble of big rocks, from which we could see a sweep of country perhaps a hundred miles across. We were surrounded, as far as we could see, by high, steep, treeless mountains, the slopes strewn in many places with gray outcrops and boulders. In the farthest distances were snowcapped peaks; the snowline here is something like 18,000 feet, so these were not humble mountains. Below us we could see a truck, a black speck the size of a beetle, moving along the road; and in the immense quiet we could hear its engine. The scale of the country is enormous—grander than anything in the Rockies. And here, as nowhere I have seen in North American mountains, there were fields in every direction, in places joined edge to edge across wide slopes, in places scattered apart, tucked in singly among rocks or in draws where moisture was available. And everywhere there were people at work in the fields or herding animals. The herds were never large, but they were numerous, surprisingly numerous; if you looked carefully enough at some distant slope or ridge you would soon make out the solitary black speck of a grazing cow, or the several small specks of a little band of sheep. Somewhere below us we could hear a shepherd whistling a slow, clear melody like birdsong.

Near that place, a little later, we left the gravel road and drove along a pair of wheel tracks out the backbone of a ridge in the direction of a herder's village. We left the car beside the tracks and walked a mile and a half or two miles across the slope of a long spur to a field where potatoes were being harvested. The field belonged to one of Steve's informants, a young farmer named Marcial Pérez. On the way down we met Marcial's wife, who was sitting with two children on an outcrop of rock, watching some llamas. She greeted Steve, joined us, and walked to the potato field with us, her baby asleep on her back in a shawl or *manta* knotted around her shoulders. When we reached the field she went straight to work, digging like the others with a *lampa,* a short-handled, backward-slanting hoe, pitching the potatoes as she dug them onto a square of heavy woolen cloth. When the cloth was full one of the workers would gather the corners to make a sack, and carry the potatoes to a loading place where they were poured into bags holding about two hundred pounds, each bag being a load for a horse or a burro.

In this field, which could not have been larger than two acres, there were twenty-two people, including a few children and two babies. In sight, in and around the field, there were also nine dogs, seven pigs, twelve sheep, nineteen horses, and fifteen llamas and alpacas. The name of the field was *Tronco,* Tree Trunk.

Nothing better reveals the long human history of the Andes and the topographical intimacy of Andean agriculture than this naming of the fields. In his book on the Vilcanota Valley, Daniel W. Gade wrote that "almost every parcel [of land], no matter how small, has a name to identify it. Property is identified by the names of the individual parcels and not by surveyor's measurements which are in most cases non-existent. Every field . . . and every enclosure has a name, many of them plant names, reflecting their immediate natural environment."

This field was yielding several varieties of potatoes, all native: red, blue, and yellow potatoes, as well as a long, flat, yellow one with red stripes, which was called Cow's Tongue. Occasionally there would be an exceptionally knotty potato, known as *lumchipamundana,* "potato that makes the young bride cry" or "test of the young bride." And that name, probably, is another clue to the frugality of these people. Such a potato would not be laid aside as a curiosity or a nuisance; it would be eaten like the rest, peeled carefully so that not even the knots would be wasted.

Early in the day a small pit at the field's edge had been lined with hot stones and filled with potatoes, and covered first with straw and then with earth. This, in Quechua, is called *pachamanca,* "earthpot." When noontime came, the pit was opened, the straw laid out beside it, the roasted potatoes laid on the straw. This, with a relish of onions and hot peppers, was the noon meal, to which Steve and I were made welcome. Sixteen of us, plus the two babies, gathered in a tight circle around the opened earthpot and ate, peeling the potatoes with our fingernails.

The potatoes were excellent—much richer and drier than ours—and the company was worthy of them. The Andean Indians have a reputation for being cold, reserved, and gloomy. These were the opposite. The outsiders were made to feel at home, and during the meal there was much talk and laughter—a lot of giggling among the girls. There were bright, friendly, generous people who loved jokes. They were delighted with the outrageousness of Steve's assertion that I spoke only Quechua, no Spanish.

When the meal was over, the workers scattered out to rest a bit, taking after-dinner chews of coca. We had to go. Two boys walked back to the car with us, talking, running, shooting at birds with slings made of inner-tube rubber. They had been sent to bring back two bags of seed potatoes that Steve had brought for Marcial, their uncle. At the car Steve gave them the potatoes and several pieces of candy. They set out together through that immense landscape, carrying a bag apiece. After they had gone several hundred yards, they sat down to eat the candy, and then went on.

Later we stopped the car again and walked down into a lovely creek valley where twenty or thirty people were hard at work making freeze-dried potatoes or *chuño*. This is a complicated process, involving freezing, soaking, and drying. And a lot of work. The end product is almost chalk white, dry, light in weight, and can be kept in storage much longer than ordinary potatoes. The work was communal and convivial, in feeling a good deal like a hog-killing at home. It was a clear, cool day like our best early winter days—ideal hog-killing weather. The short grass of the creek bottom was stippled with potatoes spread out to freeze or dry.

By eight o'clock the next morning we had finished breakfast and were on the road again, having supplied ourselves with a picnic lunch of rolls, canned tuna, bananas, and cookies. We drove up the valley, and then from the village and convent of Santa Rosa de Ocampa we turned northeast on another narrow gravel road. In the fields along the way people were harvesting, threshing, plowing with oxen and with the foot plow. Near the village of La Libertad we stopped to collect some seed of bitter lupine at the edge of a field. There we met Steve's friend Raimondo, who was coming down the road on his motorcycle and decided to go along with us.

We went on up past the diversified farms into the zone where only potatoes were planted. Their limit here is about 13,000 feet. At that altitude the farmers allow the potato fields to lie fallow seven years between crops. Above the potato zone is the zone of high pasture, where you see mostly llamas and alpacas, a few sheep. The people of this zone are herders, and one sees, scattered over the slopes, their isolated houses and stone corrals.

Above Chicche, we crossed a pass at nearly 15,000 feet,

which gave us another hundred-mile view with snowy peaks in the distance. The winter colors of the land-scape—subdued browns and yellows, pale greens—were clear in the morning sunlight. The road went by a lake lying in the broad, shallow cup of the pass, and then descended a narrow rocky gorge. That is where we had our wreck. Going down around a blind hairpin curve, we met a truck coming up. It happened too fast to scare us much, and did little damage—a single ruffle in the front bumper of our Volkswagen. One of my superstitions is that we all have a quota of wrecks that we have to happen to, and I am always relieved to get past one of mine unscratched. I felt like this was my wreck for this trip, and I was right—though some of the road we had ahead of us would make me wonder.

The road went on down along a steep, clear stream, a stairway of lovely pools and rapids stepping down through tumbles of massive gray boulders. Even in the roughest parts of this gorge were little fields, some only eight or ten feet square, nestled among the stones. There were fields steeper than barn roofs ending in sheer cliffs, where a fall would be death. (Andean farmers do occasionally fall out of their fields.) And we continued to see herders with flocks of sheep and llamas.

The valley gradually deepened between higher and higher slopes that grew greener as we descended toward the dense "cloud forest" of the *ceja de montaña* (eyebrow of the jungle). Some of the slopes are so steep that they are left in trees; on one forested, rocky, nearly vertical face we saw the silver plume of a waterfall. But some slopes nearly as steep were almost entirely farmed, and a few slash-and-burn patches had been opened even on the forested slopes; in a place or two we could see smoke rising from burning brush heaps. This is a terrain as rugged and difficult, I suppose, as any in the world, but lying under a varicolored quilt of little fields, exquisitely farmed. The country here was green, for on this side of the sierra the rains come all year round.

A little beyond the village of Comas, we stopped to watch a man plowing with an ox team a field almost too steep for them to stand on. His family was with him, breaking up the clods and leveling the field where the plow had passed. He worked back and forth across the slope, from bottom to top, with no fuss or trouble, and with surprising speed, the furrow turning downward by force of gravity (it was not a moldboard plow), the loosened clods sometimes rolling several feet.

Population pressure—the hunger of those millions in Lima—is forcing new fields into production up here. You can tell the newer fields from the older by the relative thinness of their hedgerows. The new fields appeared to be somewhat steeper than the older ones, which are plenty steep themselves, but I saw remarkably few signs of erosion. Working this steep ground produces what Steve calls "quasi terraces" or "linchets," even where terraces have not been intentionally built. As the plows and hoes loosen the ground, it shifts downward, leaving a nearly vertical wall above, and slowly piling up and leveling at the bottom. At the lower edge there may be a rim of stones picked up off the field, and here a hedge of wild shrubs and other plants is allowed to grow.

Near this field we first encountered a species of wild tobacco, a purplish, narrow-leaved, tall plant with fragrant yellow flowers, suckering at the axils of the leaves like the domesticated plants. This species was said to be too bitter to smoke.

We climbed a long stretch of road that was just a shelf along the almost sheer mountain wall, hairpinning into deep coves, taking maybe three miles to go one. High up, we stopped on the roadside for lunch. As we were getting our stuff out of the car, two men came down the road on bicycles, one of them whistling "Blowing in the Wind"— which, according to Raimondo, happens also to be the tune of a Protestant hymn sung in this country. And then a large gray-backed hawk sailed along the bluff below us, not fifty feet away.

Back the way we had come we could look up a long *quebrada* or steep, narrow draw to a black peak covered with snow. The other way the main valley opened down and down into the distance. We were perhaps a thousand feet above the stream, perhaps three thousand below the nearest mountaintops. Big gray clouds hovered around the highest crests, but it was sunny where we were. The slopes came down to the stream as steeply, it seemed, as they possibly could and still stand, leaving almost no flat land at the bottom. We ate, talked, looked at the country, watched the hawks. Aside from our own voices, the only sound was that of the distant stream.

A man, a woman, and two burros passed down the road, the two burros and the woman loaded with firewood.

The slope across from us, backed by another much higher one, was intensively farmed, but from where we watched it at first seemed deserted. Then, using binoculars, we saw a little party of workers digging potatoes in one of the fields, and gradually we discovered more and more people at work here and there over the whole face of the slope. While we watched, several of the groups gathered into the shelter of terraces or rocks. While we ate our lunch, they ate theirs.

Writing these notes three weeks later in Kentucky, I am aware how much the memory of that day has already faded. In my mind's eye I still keep a clear enough picture of the scene. But that is not what I am talking about. What seemed so alluring and charming then, and seems so hard to recover now, is a live sense of contrasting scales. The scale of that landscape is immense, so large as to constantly upset a stranger's judgment of distance and proportion; but within that immensity the Andean peasants practice an agriculture as small in scale, probably, as any in the world. Perched on the narrow ledge of that road, we were watching people working at least a mile away in fields the size of kitchen gardens, known with the intimacy of the lifetimes not just of individuals but of families—a knowledge centuries old.

It is in lingering over this contrast between the panoramic and the intimate that one begins to understand how farming and farmland have survived in the Andes for so long. For those fields hold their soil on those slopes, first of all, by being little. By being little they protect themselves against erosion, but their smallness also permits attention to be focused accurately and competently on details. This is a way of farming that has obviously had to proceed by small considerations. It has had to consider dirt by the handful. Every seed and stem and stone has been subjected to the consideration of touch—picked up, weighed in the hand, and laid down. The prime characteristic of the native, pre-Spanish agriculture was its concentration upon each individual plant, which accounts in large part for the great varietal diversity of the native crops. The Spanish agriculture, by contrast, focused not on the individual plant but on the field. It is the difference, still observable in the present practice of the farmers, between the hoe and the plow, between the potato field and the wheat field. The ox-drawn plow of the Mediterranean, old as it is, is an innovation here.

On our way back to Huancayo we passed a field where such a plow was in use, and I got out to watch. The plow, with the exception of a metal share, is built entirely of wood. There is a long horizontal member to which the share is attached. Into this are mortised a single straight perpendicular handle and a long curving tongue that fastens to the yoke. There is no moldboard; this is not a turning plow. The yoke is bound with rope to the horns of the team, here—and, I believe, usually—a pair of bulls. Such a yoke seemed to me much less wieldy and comfortable than the shoulder yoke. This field had already been plowed once, and the new furrows crossed the earlier ones at right angles. While I watched, a cold shower fell. The sky was now full of clouds and little rainstorms. As we went on, we saw a rainbow arching over the valley.

Steve talked of the difficulty of finding out about methods and reasons from these farmers. They do as they have done, as their ancestors did before them. The methods and reasons are assuredly complex—this is an agriculture of extraordinary craftsmanship and ecological intelligence—but they were worked out over a long time, long ago; learned so well, one might say, that they are forgotten. It seems to me that this is probably the only kind of culture that works: thought sufficiently complex, but submerged or embodied in traditional acts. It is at least as unconscious as it is conscious—and so is available to all levels of intelligence. Two people, one highly intelligent, the other unintelligent, will work fields on the same slope, and both will farm well, keeping the ways that keep the land. You can look at a whole mountainside covered with these little farms and not see anything egregiously wasteful or stupid. Not so with us. With us, it grows harder and harder even for intelligent people to behave intelligently, and the unintelligent are condemned to a stupidity probably unknown in traditional cultures.

I am somewhat troubled by the assumption in the preceding paragraph that the agriculture of the Andes achieved its present form "long ago." I am not sure that

that is so, and yet it is my impression that for many years, perhaps since the Spanish conquest, innovations have not come from within the peasant culture or from the contacts between neighboring peasant cultures, but have been imposed from the outside. Certainly, there was a time when the native culture was astir with profound originations, when works of genius were accomplished in the development of cropping systems and systems of soil conservation, in plant and animal domestication and selection, in the invention and refinement of tools. But it seems equally certain that this native genius must have been profoundly disturbed and encumbered after 1532 by the need to deal with a series of alien innovations: the Conquest itself; the implantation of European crops, technology, and techniques; the *encomienda* system and the oppressive agricultural quotas and tributes of the colonial period; and now the pressures of urbanization, industrialization, and population growth, as well as the technology, techniques, and economics of "agribusiness." It is true that the peasants exercised a certain amount of selectivity in their adoption of Old World plants, rejecting turnips, for instance, in favor of their own root crops. Whether or not local tradition is conscious or confident enough to stand in its own defense against New World "agribusiness" remains to be seen. The capitulation of North American agricultural tradition (such as it was) is not a reassuring precedent.

By this time I had looked carefully at a good many fields and passingly at hundreds, talking always with Steve about what I saw. I had begun to be impressed with the way erosion control is built into the patterns of this farming, as one of its dominant themes. I started a list of the various means of erosion control, to which I kept adding as I went along:

1. Smallness of scale. The small fields seem often to correspond to a single feature or facet of the terrain. Where there is a slight change in gradient or exposure, one field ends and another begins.

2. Terracing.

3. Fallowing, which keeps some fields unbroken for as many as six out of every seven years.

4. Proper timing. The ground is worked and planted mainly during the dry season. When the fields are most vulnerable to rainfall, no rain is falling. (On the eastern slope of the sierra, where there is no dry season, the fields can be kept almost continuously under growth of some sort, which may afford the opposite kind of protection.)

5. Polyculture. So far as I remember, the fields I saw were all planted to single crops, but Daniel Gade found polyculture "deeply engrained in peasant custom" in the Vilcanota Valley. And he points out that hoe agriculture permits a greater diversity of species in the same field than does plow agriculture. The advantage of polyculture in erosion control is in the overlapping of the life cycles of the various plants, or in the provision of cover at different levels, as when low-growing crops are grown under trees.

6. Downhill rows, which let the water out quickly with minimal damage.

7. Weeding only after the crops are well established.

8. Planting in sod, breaking the ground only after the crop has started to grow. This method requires hand digging. It seems likely to me that the ground that is hand dug may be less vulnerable to erosion than ground that is plowed, because hand work breaks up the sod less. Steve says that in the hand-dug fields the sod is merely cut and turned over, not pulverized.

9. The return of organic wastes to the soil: crop residues, animal manure, ashes—everything but human manure, which I understood is recycled by the hogs and so is returned to the fields as hog manure. The primary aim, here, is of course not erosion control, but erosion control is an important by-product. A soil that is fertile and has enough humus absorbs water better, holds together better, and covers itself better with vegetation than a depleted soil.

After breakfast in Huancayo Thursday morning, waiting for the bank to open, we walked through the market, which seemed a busy, crowded place to me, though Steve said that it was unusually quiet. Booths selling various kinds of produce and merchandise—meat, vegetables, fruit, fish, live poultry and guinea pigs, weavings, clothing, pastries, etc.—crowd the large hall which is the market proper, and spill out along the nearby streets. All the groceries seem of good quality and are appetizing enough, until you look down at the floor, which is a mess. Some people sell their merchandise out of tiny booths; some bring their stock of dry goods in a suitcase or two; some spread their wares on a few square feet of pavement; some who own a mechanical stitcher and a few tools set up in the open street as shoe repairmen.

From the market we went to the bank. I had a $20 bill that I needed to exchange for *soles,* an operation that I expected would take two or three minutes, but which took perhaps three-quarters of an hour. After standing in line at a window, we gave the money to a teller, who filled out a form with four carbon copies, each copy of which had to be stamped with four rubber stamps, each stamping of which had to be cancelled with a pen scribble. The teller gave the bill and the forms to an administrator sitting at a desk behind him, who checked and approved the forms, and returned them to the teller, who gave them to us. We then stood in another line before another teller, who at some leisure counted out the *soles*.

All this, along with my bafflement, reminded Steve of one of his and Peggy's entrances into the country, when they and the immigration authorities filled out elaborate forms, which were then solemnly handstamped, scribbled on, and deposited in a wastebasket.

And later he told me of his triumph over the Post Office. He and Peggy had received a Christmas package, on which they were notified that they would have to pay an enormous duty. Steve went down and told the several officials that he could not afford the duty, but he would gladly pay the postage to return the package to the relative who had sent it. No indeed, they said. He would either have to pay the duty or forfeit the package. The idea, obviously, was to make it impossible for him to have it so that *they* could have it. "Well, may I see it a moment?" Steve asked. They handed it over. Whereupon he threw it to the floor, stomped it flat, and walked out.

When we got free of the bank we drove northward toward the upper end of the valley, where the farming is more commercial than in other places we had been. The fields are larger there in the bottomland, and some of the work, such as plowing and the harvesting of small grains, is done mechanically—Steve thinks about 5 percent of it. Most of the work, however, is still done by hand or with ox teams, the produce carried from the field by pack animals, in the peasant fashion.

We stopped at a field where several women were cutting quinoa with sickles, and four men, under the eye of a young supervisor, were threshing. A large tarpaulin was spread on the ground, the sheaves of quinoa laid on it, and the seed beaten out with flails. The flails were simply staves or poles six or so feet long, slightly curved. The men used the flails with an energy that surprised me, beginning the stroke far back, as with an ax or sledgehammer. To winnow the grain—and, I believe, to shovel it into bags—they used a broad paddle about three feet long. While the men worked, a little girl no more than six years old stood by with a baby tied in a shawl or *manta* on her back.

And as everywhere, there were dogs. These were vicious, as they frequently are. The supervisor was almost constantly occupied in keeping them from attacking us.

The uncut quinoa stood about three feet high, planted in rows. The stalks, with heavy seedheads, had ripened almost to whiteness. This crop is never harvested or threshed mechanically; a machine would waste too many of the small seeds.

Quinoa is a native plant, which, so far as I saw and according to what I have read, is no longer widely cultivated. Apparently it is being replaced by barley and wheat, which is unfortunate, for, according to Professor Gade, this crop has several advantages: "few pests, high food value, a variety of uses for most of the plant, and ability to produce despite low rainfall and temperatures and poor soil conditions." Quinoa is richer in protein than the European small grains, and its protein is of exceptional quality.

We drove on to Laguna Paca and its lovely village, where hollyhocks, snapdragons, and other flowers were in bloom. For lunch we had a trout and a beer apiece on the sunny veranda of a small restaurant beside the lake, while we looked across the water at the mountains.

As we were leaving Paca we saw a small, self-propelled Massey-Ferguson combine with what looked to be about a five-foot head—anyhow, it fit with room to spare in one lane of the road. I wondered why such small combines are not for sale at home. In a Peruvian valley a few of these machines would produce an agricultural revolution, profoundly disruptive of established economies and ways. At home, where ordinary combines are now too large to drive through a twelve-foot gate, and where operators can no longer afford to harvest little fields, such machines would help the small farmers to survive. This is a perfect paradigm of "agribusiness" economics (and politics). No new machine is ever introduced that will help the existing community to survive. The available technology must be barely within financial reach of the biggest operators—that is, it must be as socially destructive as economically profitable. The manufacturers profit most from increases of scale, not community stability.

We climbed back up toward the pass above Chicche, which we had crossed the day before on our way to Comas. As we went up, Steve stopped to leave a gift of seed potatoes for another of his friends. We had to hold off a pair of dogs by throwing rocks until the farmer's son Luis came out and put them in their place. (I managed to hit one of these dogs a good solid lick with a rather sizable rock, which gave me the deep satisfaction that invariably comes with the fulfillment of justice.) Luis and his cousin, a boy about nine years old, decided to make the rest of the trip with us, for it was Luis's uncle we were going to visit.

We crossed the pass and drove to the far end of the lake. As we left the car a pair of large black and white geese rose from the shallows and flew away over the water. We set out toward some shepherds' houses perhaps two miles away.

As we walked, because of the altitude, which was almost 15,000 feet, my cap seemed to draw tighter and tighter around my head, and the air seemed never to fill my lungs fully enough. But it was an exhilarating walk nevertheless, with the lake several hundred feet below us and the country grand around us in the sunshine. Big clouds hovered on the peaks. We were walking over tundra, and everywhere underfoot were scattered the exquisite, tiny blooms of alpine flowers: yellow ones, tuberous ones mainly red but ranging from yellow to purple, blue ones with yellow centers like forget-me-nots, white ones with yellow centers like daisies. There was something rare and moving in the contrast between the great heights and distances and those perfect little flowers. And though the air was painfully thin, one was always more or less aware of its purity. The light too was pure and strong, with the surprising clarity of the clearest water.

After we had gone perhaps a mile, a man with two dogs suddenly appeared over the ridgetop above us, carrying a loaded *manta* and a drop spindle. This was the shepherd we had come to see, Nestor Figeroa. He came down to us and greeted us with handshakes. He was a weathered man of about fifty, his speech deliberate and emphatic, his eyes direct under the turned-down brim of an old felt hat. It was a countenance shaped and steadied by solitude and by distance. This was a man, one immediately felt, who would say exactly what he intended, never more nor less. Among his people, Nestor is a man of considerable means. He owns many head of livestock—sheep, llamas, and cattle—and he has a house and a wife down in the village. But he is "a man of the *puna*," Steve says, and he prefers to be up here.

The five of us went on again, passing through a herd of grazing llamas. By now the little boy had taken charge of my binoculars and had begun to lag behind with Luis, looking at things.

Some comment being made about the flowers, Nestor immediately knelt down near one of the little blue ones. He rapped with his knuckles on the hummocky tundra beside it. The blossom began to close. In a few minutes it had drawn its petals together tightly. Its Quechua name, Nestor said, is *manalika wayta*, "flower that is embarrassed."

Nestor left his *manta* and spindle beside a rock, to be picked up on another of his rounds later in the day, and took us past several grazing cattle to a shelf of level ground above the lake, where an old man and two teen-age girls were making *chuño* of the bitter potatoes that had been packed up on llamas from the potato fields 2000 feet below. Among the boulders and gray outcrops on the slope above us were stone corrals and two stone huts with high-peaked, thatched roofs. Beside the rectangular corrals presently in use were the vestigial stone circles of ancient ones, marked with green, gray, and orange lichens. Steve gave cigarettes to the two men, and we sat there in the wind and the brilliance, visiting. The black and white geese flew again over the lake.

The two nearby huts, it turned out, belonged to the old man. We were now going on to Nestor's place. As we got up to leave, the two girls, who were going to ride back down the mountain with us, set off around the shore of the lake toward the car. Steve, the two boys, and I followed Nestor up over a ridge and through a narrow notch in the rock. Beyond the notch, the country opened again, and we came up into the smoke of a dung fire in front of Nestor's huts. The huts of the shepherds are always in pairs—one for storage, one to live in. They are tidy, windowless little buildings fitted snugly into the landscape, looking at once natural and human. They consist simply of a mud-chinked stone wall about three feet high, with a single low door and a high-pitched, four-sided thatch roof.

Nestor disappeared into the living hut and came out with a sheepskin, which he spread out on a ledge of rock beside the storage hut. He motioned that I was to sit on the sheepskin. He and Steve are friends, *compadres*. I was a guest. His gesture was consummately gracious. I thanked him and sat down. He disappeared again, and returned with a small, much worn enamelware pan of roasted potatoes, apologizing, as Steve would tell me later, for having no more to offer.

I was more than satisfied with what he had to offer. I sat with the two boys on what may be the best front porch in the world, certainly one of the highest. While Steve and Nestor talked, I ate potatoes and looked at the hills and the lake, the rocks and the clouds. And then I watched Nestor instruct Steve in the use of a llama-wool sling that he was sending as a gift to Steve's son, Jason. Jason had paid Nestor a visit once, and Nestor remembered him with affection.

Across the lake, just moving specks in the distance, were several herds of sheep and llamas. As I watched I slowly made them out. Those high pastures are stark and austere, vast in their silences and distances, roughly shaped, naked to the wind and rain and snow. Beautiful as they are in sunlight, in bad weather they must be singularly harsh and difficult. To a shepherd, alone in his hut at night, or crouched in rain in the lee of a rock, the solitude must be profound. And yet this is a human landscape, in which one never has to look very far or very long to see dwellings or corrals or herds or people.

And even there the responsibilities of agriculture are remembered. Beside the fire was a pile of ashes and beside the corral a pile of manure—both to be taken down to the fields.

As we started to leave I asked if I might see the inside of the dwelling hut, and was assured that I was welcome to do so—though this again caused Nestor some embarrassment. He apparently felt that the presence of a guest in his house called for something special in the way of food or drink, but he had not known that we were coming and had nothing prepared.

The door was so low that I entered with difficulty even on my hands and knees, and inside I could not stand erect. The combination of my height and my speechlessness made me feel more than a little awkward in this man's hospitality and in his house. And yet few houses that I have been in have interested me as much. None that I have been in has so precisely—so perfectly without lack or excess—answered the need of its dweller. In the righthand corner beside the door was a domed adobe fireplace with an opening perhaps eight inches wide. In the opposite corner was a small pile of dried llama and cow dung to be used for fuel. There was no flue, the smoke escaping through the thatch which was thoroughly blackened. At the far end of the room was the bed, an earthen platform covered with straw and then with blankets. Huddled together on the sleeping platform were four or five black and white guinea pigs, the ancient household animals of the Andes, whose low cry sounds very much like "oink," but in Quechua it is the sound of their name: *cuy.* Above the bed, on a pole fastened between the slopes of the roof, hung several fleeces used for bedclothes. Strips of smoked meat hung from a stretched rope. On a stone above the fireplace sat a tiny oil lamp. That was all. That was enough.

What made it enough, of course, was not just its physical properties, but the culture of the man who lived in it: his inheritance of many generations of familiarity with the country, and of the skills that have enabled so many generations to live there. To a stranger, lacking this culture, this house would be inadequate and intolerably lonely. But so, lacking an appropriate culture, is *any* place—or so it will sooner or later get to be.

On the way back to the car, one of the boys suddenly stopped us and pointed across the lake. We looked and saw a flock of sheep, and then a shepherdess in a red skirt running across a swale between two stone outcrops. She was almost too far away to see, but Luis recognized her and told us who she was.

We got up at five o'clock on Friday, and breakfasted on sandwiches bought the night before. Our plan was to make an early start, climb up into the snow on a saddle of Mt. Huatapallana, and get back to town before noon.

Before daylight we were on our way, climbing out of the valley through yet another steep narrow draw—a very pretty one, with a lot of trees. The sky was overcast and gloomy, and we drove through patches of drizzle, hoping for clear weather higher up.

We passed an old *hacienda,* one of the many manor estates established under Spanish rule, this one apparently consisting mostly of grazing lands. The old owners of the *haciendas* have been dispossessed by the present government, which now manages these estates. Grazing is controlled on the *hacienda* lands by some system of pasture management, and the pastures here were more abundant than others we had seen.

Above the *hacienda,* the drizzle turned to snow, whitening the ground. We passed a herd of twenty-five horses being driven up to pasture by two horsemen in ponchos, looking cold with the snow melting on their hats and shoulders. They were riding very smooth-gaited horses. Everywhere I saw them, the Andean horses were small, but extremely tough, capable of carrying a grown man at a gallop over the mountainsides.

Later we passed a man and a boy with a train of llamas. They had crossed the mountains with potatoes to trade for corn, which they were now bringing home. The loads, the man said, weighed about 130 pounds. They had been traveling for eight days.

When we reached the pass where we had intended to

leave the car, it was still snowing and we had to give up our climb. There was already enough snow to make the footing difficult, visibility was poor, and there seemed to be some danger that the snow would get deep enough to slicken the road. We had hoped to go up to the snow, but the snow had come down to us.

As a substitute for mountain climbing, we went to Cochas Chico and Cochas Grande, two villages where beautiful carving is done on gourds. We visited several houses, and at one bought several gourds to take home. These are neat, pretty villages, predominantly Protestant, with many eucalyptus trees and irrigated gardens of vegetables and flowers. At Cochas Grande an old woman sat beside the road with a galvanized bucket in which she was cleaning the feet and head of a sheep.

On Saturday we were to return to Lima. We got up at four o'clock, breakfasted again on sandwiches, and began the trip, taking along Matt Garr, a Jesuit priest from Louisville, Kentucky. Matt is thirty-three, bright and likable, full of information, intelligence, and humor.

According to Matt the ideal population on the mountain pastures is one sheep per hectare; at present, however, there are sometimes as many as twenty. These figures, with the well-managed *hacienda* pastures we had seen the day before, seemed pretty clearly to indicate that much of this pasture is overgrazed. (Much that we saw *looked* overgrazed to me, but not seriously so.) This, as I understand, is caused by population pressure. The surplus population of the mountains migrates to Lima, which takes certain pressures off the mountains. But in Lima people have to eat and there they do not produce food.

As population growth and movement increase, along with the stresses of industrialization and modernization, suicide becomes one of the means of escape. In the highlands, people kill themselves by drinking insecticide—one of the more remarkable applications of "agri-industrial" technology.

The sort of technology that is most needed, Matt said, is the sort that does not cost anything—mainly certain improvements in know-how. For instance, great good might be done by some method of breeding the ewes so that they will lamb when the grass is best. At present, the rams run with the ewes continuously, and so the lambs come at the wrong time. Matt confirmed my observation that male animals are usually not castrated. The quality of the livestock could be quickly improved by keeping only the best males for breeders.

Matt told us this story about the modernization of the peasants. A nun, who taught school first in the mountains and then in the *barriadas* in Lima, would ask the children to draw a picture of "the most important person in your life." In the mountains, the boys would draw their fathers, the girls their mothers. In Lima, many of them would draw a picture of the television set.

And he gave us a rule for driving in Lima: it is safer to run a red light than a green one. If you run a red light, you will slow down and look both ways. If you run a green light, you will not look, and will be hit by somebody running a red light.

As we came into Lima, Steve pointed out a small suburb of slum houses built on the tiers of an Inca pyramid. We turned in our car, parted with Matt, and took a taxi to the apartment where the Brushes were staying.

During lunch, young Jason said of Nestor Figeroa: "He's somebody who lives way up." I agreed.

Now that the Huancayo journey was behind me, I could see clearly my indebtedness to Steve. In so short a time I had had the benefit of his long study of Peru and of his year of work around Huancayo. Not to mention his knowledge of Spanish and his familiarity with the people. He has made himself a friend to the farmers whose fields he studied. He likes and respects them, which carries him far beyond the role of "objective observer," and appropriately complicates his insights and his tasks. This makes him, so far as I am concerned, many times more trustworthy than any "detached" scholar.

And it makes him an ideal guide. Thanks to him, I had seen in six days more than I could have seen on my own in six months.

The farming he showed me seems exemplary in many ways. Whereas our agriculture is focused almost exclusively on short-term production, the traditional Andean agriculture is focused on long-term maintenance of the sources of production. The themes of our agriculture are volume, speed, man-hour efficiency. The themes of Andean agriculture are frugality, care, security in diversity, ecological sensitivity, correctness of scale.

Some people may be inclined to explain the frugality of these farmers by their poverty, and thus dismiss them. But that is too simple. One does not take care of things by needing to do so. One does it by knowing how. The Andean farmers are worthy of our interest and study because they have known how to take care of their land and maintain its fertility over many hundreds of years. The better we understand how they have done it, the less likely we will be to assume—to our great danger—that it can be done simply or easily.

I was also impressed by the industry of the Andean people. They work both well and hard. They are far indeed from the stereotype of the "idle peasant." In comparison, American workers—with their short hours, long weekends, slowdowns, and shoddy work—are the ones who look shiftless.

The most unattractive thing about the Peruvians that I

saw was their pervasive distrust of each other, manifested in the cities by an apparent obsession with walls and locks, and in the country by mean dogs and by guard huts or watchtowers in the fields. This distrust may also explain the bureaucratic redundancy and the lack of courtesy in traffic.

I spent Sunday in Lima, mostly resting, and on Monday morning flew to Cuzco. At the airport I sat by a couple from Atlanta. The man wanted to know what I was doing in Peru. I told him I was looking at farms in the Andes.

"You a farm equipment salesman or something?"

I explained my interest, mentioning among other things the loss of topsoil by erosion in such places as Iowa, where one bushel of corn now entails the loss of five or six bushels of topsoil. "I can't believe we'll ever run out of land," he said. "I didn't know you could use up soil." We were mutually amazed.

From Lima we first flew over desert mountains without any sign of life, farmed only in the narrow valleys where irrigation is possible. And then a few little unirrigated fields appeared on the slopes, and then more and more fields, lakes, the rectangles and circles of stone corrals. One deep, steep valley after another opened below us, with never more than a narrow ribbon of flat land at the bottom. This is what Peru must depend on to feed itself. From the air you can see unmistakably how fragile this agriculture is, how delicately it has been kept poised for so long upon the steep slopes under heavy rains—and how dependent it necessarily is upon the skill, the culture, of the farmers. If, under the use of industrial technology and industrial economics, the topsoil of Iowa will be gone by the year 2050, how long would the topsoil of the Andes last?

In the Andes, the questions of scale and proportion are

clearly paramount. The fields *have* to be the right size; to make them too big would be to destroy them. And there *has* to be a correct proportion between the number of farmers and the acreage of farmed land. If the number of farmers should be reduced by too much, as by the introduction of industrial technology and economics, then priority would shift to production, to the neglect of maintenance, and the land would be lost. Too much food can produce starvation as easily as too little.

What I was thinking, then, looking down at the little fields of the Andes, was that the most interesting, crucial, difficult questions of agriculture are questions of propriety. What is the proper size for a farm for one family in a given place? What is the proper size for a field, given a particular slope, climate, soil type, and drainage? What is the appropriate crop for this field? What is the appropriate kind and scale of technology? Andean agriculture is a success—has lasted thousands of years on extremely difficult terrain—because it has so far answered such questions correctly.

These questions are as critical for us as they are for the people of the Andes, as our troubles with erosion, soil compaction, etc., plainly show. The difference is that we farm, generally, on flatter land, and for us the questions have not been so obvious. So far. We have had the luxury of pretending that the questions do not exist, that there are no problems of propriety or proportion and no limits to scale.

As the plane flew on inland, there was a noticeable improvement in the pasture, and the fields became more numerous. There began to be trees, probably eucalyptus, in the water courses and on the slopes. Cuzco is in a comparatively wide valley, surrounded by beautifully kept fields.

That afternoon I spent a lot of time standing in lines to change money, make reservations, etc. But I also walked around the city, looking at Inca remains and at churches.

At its best, Inca stone masonry is as excellent as the books and pictures lead one to suppose. The workmanship could hardly be surpassed. The Spanish churches are obscured, for me, by the violence that justified itself by them. I cannot help seeing them as symbols of the conflict between the meaning of Christianity and the uses it has been put to. Conquistador Christianity was, at least, a contradiction in terms. The Cathedral is overwrought with silver and gold, gaudy as all the work of guilty conscience. Of course, the Incas too were conquerors, geniuses of empire and domination, their religion as much a state religion and as serviceable to power as the Christianity of Pizarro. Spanish gold was first Inca gold. The Church of Santo Domingo is built on the foundation of the Inca Temple of the Sun. Both, I think, were erected on the peasant culture (and agriculture) that preceded, supported, and suffered them.

I came to Cuzco, after my week at Huancayo, at approximately the same elevation, expecting that I would be used to the altitude. And so I put in a fairly strenuous half day, and ate two sizable meals in the bargain. As it turned out, I was far too confident. Monday night my heart sped up to a hundred or so beats a minute, making rest at once necessary and impossible. On the advice of a doctor, I took some medicine; and, having postponed a planned trip to Machu Picchu, I spent Tuesday morning resting.

In the afternoon I hired two boys in an old car to take me to the village of Chinchero. This is an ancient site, the present village being built on Inca ruins, and surrounded by Inca stone-faced terraces. And it is extraordinarily beautiful. It is surrounded by open, gently rolling country— excellent farmland—that reminded me of parts of the Great Plains, with snow-capped mountains off in the distance.

With one of the boys, who spoke a few words of English,

I walked out beyond the village to a large stone outcrop overlooking a little valley. An entranceway with stairs had been tunneled right up through the heart of the rock. Up on the rock, seats and steps had been carved in a number of places. There was no obvious clue as to what the reason for all this work might have been—unless to provide a fine place to sit and look at the country, which it certainly was.

Below us, along the valley floor, were little stone-fenced potato fields, and out to the right the walled terraces stepping down the slope. The stonework in these walls is excellent, very nearly as good as that in Cuzco, and it provides a good insight into the agricultural thinking of the Incas. Whatever one's opinion of the organizational genius of those people—my opinion, though respectful, is hardly adulatory—it did make it possible for them to confront head-on the problem of duration, something our own planners, in spite of their "futuristic" obsessions, have not been able to do. These walled terraces testify that the Incas aimed at a *permanent* agriculture, and that aim accounts for the excellence of the workmanship.

Stock was grazing in the bottom and along the sides of the valley. Two women were sitting at ease by the stream, their wash, of many bright colors, spread to dry on the grass and bushes around them. The farmers were making *chuño,* threshing oats, broad beans, quinoa, barley. The oats here were particularly fine, standing four feet tall in the shocks. Around us, we could hear children calling, burros braying, cattle bawling—and over and around these sounds the deep stillness of the mountains.

On this drive, as on some earlier ones, I saw a good deal of what appeared to be old erosion caused by cropping or overgrazing—though, as usual, very little in fields now in use. I could not be sure of the causes of the old washes, but they seemed to occur in unnatural places or to go down the slopes at unnatural angles. If this is agricultural erosion, can it have been caused under the *encomienda* system of the Spanish, which imposed heavy production quotas on the native farmers? It seems to me that careful reconstructions of the human history of some of these slopes would be extremely useful.

I also saw, for the first time, a fair-sized tractor working ground on a hillside. So it can happen.

There is now a big fertilizer plant at Cachimayo, between Chinchero and Cuzco. Thinking of what that might mean, it occurred to me that the natural limits of the fertility of these steep fields may be their greatest protection, because the natural limits impose the fallow system. Chemical fertilizer will break that system, enabling crops to be grown—and the land thus exposed to erosion—every year. Similarly, the introduction of tractors will tend to enlarge the fields, destroying the protections of proper scale.

As far as its purposes were concerned, my trip ended at Chinchero. That evening I still had my old symptoms, plus a new headache. I had got too tired to throw off the effects of the altitude, and those effects were keeping me from resting. I needed to go home, and home is where I went. When I got there, the first raspberries were ripe.

2

Three Ways of Farming
in the Southwest
(1979)

April 6, 1979
I arrived in Tucson at 10:19 P.M.—12:19 Kentucky time, a difference I had failed to consider. Gary Nabhan and Karen Reichhardt, my hosts, met me at the airport, and we drove into town to watch a part of the Yaqui Easter observance.

This turned out to be a blend of Spanish Catholicism and Indian ceremonialism—masked dancers in a double line leading to the church door, people entering in pairs to kneel and touch the images, women's voices singing old Spanish hymns. We stood around and watched for perhaps an hour. To a newcomer it seemed inexpressibly strange, taking me farther from home than the previous five hours in airports and airplanes.

April 7, 1979
My first job of this trip was to take part in a "Hunger Conference" at the University of Arizona. The panel of which I was a member, "Crucial Hunger Problems and Solutions," convened at something after 11:00 A.M., following various

47

opening remarks and a short keynote speech by Representative Udall.

Our discussion was conducted in the atmosphere of un-
reality that descends upon most academic events. The remarks ranged from optimistic (the world is winning the
battle against hunger) to pessimistic (American technology
may not be the answer). Starvation was seen variously as a
psychological problem, a political problem, an organizational problem. Agriculture was apparently considered a
rather peripheral concern; specific agricultural problems
—technology, soil erosion, etc.—were little discussed. A
lady in the audience read a quotation from somebody
who said that feeding the hungry was a matter of "political will." There was no window in the room. The light was
artificial. The audience, though the conference was accompanied by a fast, looked well fed. The problems seemed remote—though, as I was soon made aware, several were in
fact very close at hand. It was a performance characteristic
of the disuniversity.

The moment I remember best came when a member of
the audience asked if traditional, local systems of agriculture could not be encouraged and built upon as a way of
alleviating hunger. This question was addressed to the pan-
elist who believed that the world is winning the battle
against hunger.

His reply was representative of what is probably the
dominant school of thought on this subject. In order to re-
lieve hunger, he said, you have to have a cash economy.
People have to have money to buy food. Traditional ways
of farming rarely provide more than a subsistence. When
traditions get in the way of the growth of a cash economy,
they must be removed "by surgery." It is thus possible
within the length of a breath to go from paternalistic economics to tyrannical politics.

That question and its answer state well enough the con-
flicting themes of the remainder of my trip.

Toward the middle of the afternoon Gary, Karen, and I finished our obligations at the conference, loaded some camping supplies into the back of Gary's truck and started southwest on Highway 86 into the Sonora Desert and the Papago Reservation. We drove with the sun in our eyes, leaving behind the American city-fringe eyesores (the same in Tucson as in Louisville), and then the further spattering of subdivisions colonized by mobile homes each with its suckling car or pickup truck, and then the irrigated fields—and then, with relief, into the desert itself, where everything but the roadside trash and the road itself looked native, belonging in place.

I was feeling another kind of relief too, often felt before: the relief of moving from talk about problems into the presence of the problems themselves. In the presence of the problems intelligence encounters details. It is like stepping from slippery footing onto dry rock. The relief is physical. And it is hopeful too, for it is in the presence of the problems that their solutions will be found. Solutions have perhaps the most furtive habits of any creatures; they reveal themselves very hesitantly in artificial light, and never enter air-conditioned rooms.

My family and I drove eastward on this same road ten years ago, in March of 1969, and ever since I have been hoping to come back. Why the Sonora Desert should appeal so strongly to me is a mystery. My geographical ideals were formed by bluegrass pasture and deciduous woodland, and so I hardly qualify as a desert lover. That is, I don't love deserts in general. Most desert country that I have seen has shown me little that I wanted to linger for. But the Sonora Desert is different. I have a taste for it.

Perhaps its scale is right. It is broad and flat enough to give a continuous sense of largeness and distance. But it is not *too* broad and flat, for there are always mountains in the distance, and frequently there are hills or outcrops nearer at hand.

Perhaps it is the saguaro and cholla cacti. There are other cacti in this desert, but these give it its characteristic look and tone. The tall, branching saguaros have an austere dignity that no tree can equal. There is something almost human in their anatomy and stance, and yet something profoundly unhuman in their stolidity and stillness. And the spindly cholla with their crooked, bristling arms are as formidable as a mob. The thought of backing into one of them is almost too much to bear. And yet there is something gentle and gentling about this plant. Its clustered spines, especially on the recent growth, catch the light and diffuse and soften it, glow with it, giving it a quality to be found no place else.

Over everything is the desert silence. You don't have to go far from the road to hear it. It is a prevailing, enormous, patient silence, not even dented by the small sounds that occur in it. It is primordial, here before the beginning. The world occurs in it.

Before the coming of industrial civilization, the Sonora Desert was one of the most difficult human habitats in the world. There is a brief rainy season in summer, another in winter. The rainfall is light, varying from three to twelve inches; the evaporation rate is high. Except during the times of rain, the Papago country has no flowing streams. In the old days the Papago depended on a few widely scattered springs. Sometimes water had to be carried for many miles. Despite the warmth of the climate, scarcity of water made the growing seasons short. Agriculture, though cunningly adapted to the place and its conditions, was never a sure thing and was only a part of the subsistence. The people also depended on hunting and on harvesting the wild plants. They are said to have used at least 275 different species of wild edible plants, 15 percent of which were major food sources.

In response to their meager land, the Papago developed a culture that was one of the grand human achievements. It was intricately respectful of the means of life, surpassingly careful of all the possibilities of survival. Their ideal was "survival, not triumph"—so Charles Bowden writes of them in *Killing the Hidden Waters*. The Papago communities were at once austere and generous; giving and sharing were necessarily their first principles. The people needed each other too much to risk individualism and dissent. For, as Bowden says, "The man who hoarded, who saved, who said he and his blood would make it on their own . . . such a man led his kin to extinction. In the Sonoran desert a man could get an edge, but never savor control. Power came from toil and could only be stored in other human beings."

The result was paradoxical: in these almost impossible circumstances, the Papago achieved what Bowden accurately calls a "society of abundance." The poverty of our own "affluent society" never existed among them. The old-time Papago lived close to the Christian morality that we profess, and far indeed from the morality that we practice. We could do worse than put ourselves to school to them. "Having little," Bowden says, "they shared all." And they lived in harmony with their place: they "lived the net yield of the desert's hydrologic budget."

Contact with white people brought many changes, of which the most radical may have been wells drilled into the aquifers that underlay the desert. But this was a part of a government program that also included livestock production and education. Bowden notes the irony in this effort, which proposed to bring self-sufficiency to a people already marvelously self-sufficient. The opposite happened: "Education divided the tribe between those who had seen the tractor and those who had not. Stimulation of the cattle industry resulted in the ruin of the rangelands. The wells . . . cut the ground from underneath the ancient mutual dependence and sharing. A half century after the commissioners' optimistic forecast, the Papago are not respected by their white neighbors and are not self-supporting. They now have a groundwater problem, an overgrazing problem, and an economic problem. The society of abundance is gone."

Along with fossil water, government programs, and problems both local and national has come junk. The roadsides are littered with throwaway containers and other garbage, which the desert climate preserves in mint condition. We even found a pistol, which we left for somebody who might need it worse than we did. But there were also poppies and lupines, tiny cranesbills, the tall flowerheads of wild rhubarb, and many other flowers I did not know. The rains had been unusually abundant, and the desert was rich in bloom.

Like other traditional regional cultures, wherever they have come into confrontation with industrial economies, the ancient culture of the Papago is on the wane. It has been on the wane through most of this century. Whatever one may think of the opinion of the expert at the Hunger Conference, there *is* a conflict between the operations of a cash economy and traditional, local systems of agriculture. It is easier to buy your food than to grow it. It is hard to persuade a community to grow its own food once it has become available for purchase—provided that money is somehow available. It is easier to drink soft drinks and throw the containers out the window than to practice the difficult disciplines of health and frugality.*

And so the society of abundance becomes dependent on

---

*In a healthy culture, of course, personal health and frugality would not be difficult—they would not be perceived as "disciplines." They become difficult when disease and waste become normal.

a society of scarcity, consuming exhaustible resources as rapidly as possible in the conventional American Way, and leaning on the fragile props of inflated cash and government programs. And so the intricate, delicate culture so responsive to the needs of desert life is gradually replaced in the mind by modern restlessness and the desire to shop. And so the body loses its resilience and strength as its purchased diet is converted to fat.

Though the old life is no longer lived anywhere in its old coherence, fragments and vestiges survive. The old agriculture, for instance, is still practiced by a scattering of farmers, mostly old, who cling to it by habit or by preference for traditional foods. The two significant influences on this agriculture in historical times have been Spanish and modern American. The Spanish influenced it by introducing new crops and the use of work animals. The modern American influence has tended to do away with it rather than change it. And so where it survives one finds it pretty much as it has been since the absorption of the Spanish innovations around the beginning of the eighteenth century.

One of our errands on this trip was to deliver a mended plow to an old Papago woman who had no place nearer than Tucson where she could get blacksmithing done. We found the old woman at home, and spent quite a while talking with her and her son, and an elderly kinsman and neighbor. Gary had brought gifts of seeds and a tangled bunch of the hooked, thorny pods of devil's claw, which yield a fiber used in basket making. The old woman was gracious, making a large to-do over Karen and the gifts, her words as much sung as spoken. Her son was reservedly polite and informative, gazing under his cap brim into the distance as he spoke. The old man, aside from a brief approval of the work on the plow and a few bits of information, said almost nothing. Around the small house and outbuildings the ground was swept as clean as a floor.

We then drove over to the old man's place to look at his field. Nobody was there. His wife dead now, the old man lives there alone most of the time. This place too was neatly kept, the ground clean around the buildings. As seemed typical of these widely scattered dwelling places, where the people still hold to the remnants of the old desert life, this one displayed a curious mixture of artifacts: barbed wire fences and an assortment of worn horse-drawn tools, sheds walled with live ocotillo stalks, horse tracks and tire tracks.

As the sun was going down we walked out to the field which, after the first good rain of winter, had been planted in wheat and peas. The ground, Gary said, had been broken with a horse-drawn plow. The wheat was sown in the open furrow and covered by the next furrow. The peas were planted the same way, but in hills about a yard apart. Both crops were now near to bearing, and they looked well.

Before being broken, it seemed, this field must have looked much like the flat desert land lying around it. But there was a difference. The Papago word for this kind of farming is *akchin*, which means "arroyo mouth." By this method the fields are placed where the waters of the floods slow down, spread out, and soak into the soil; or where, rarely, the water table lies near the surface of the ground. In these latter places, the Papago say, "the water of the ground meets the water of the sky." A good place for a field is where the wild plants stay green during the dry seasons, where the soil stays cool, where there is a lot of dew throughout the year.

But though the plot chosen for a field must have natural advantages, it must also be improved by human care. The Papago conceive of cropland as something that people make. They "make the good earth" that is essential for farming by clearing and cultivation, by bringing in good dirt from other places, by growing and plowing in green manure crops. There is a period during February and

March that is called "the month of the short stalk planting"—that is, the planting of crops to be plowed under. Some families fertilize their fields with the leaves of the mesquite, a leguminous tree. Some turn horses into their fields after harvest or before planting to have the benefit of the manure. Not surprisingly in such a climate, it is well understood that humus increases the water-holding capacity of soil, and a high humus content is a part of the definition of "good earth."

The wheat in this farmer's field was a widely used variety known as White Sonora; the peas were Papago Peas. These, along with other such winter crops as garbanzos and lentils, were introduced by the Spanish—an example, I would say, of the best sort of innovation, for before these, the Papago had no winter crops. These made double cropping possible, which greatly increased the yield of the agricultural lands, and did so without much depleting the soil, for many of the new crops were legumes, and the "short stalk planting" continued. By now these crops are well adapted to the region and do not require much water.

The Spanish also introduced such summer crops as sorghum, watermelon, and black-eyed peas. But most of the summer crops are native: a sixty-day corn, three varieties of tepary beans, pinto and pink beans, chili peppers, striped cushaw (Papago pumpkin), butternut squash, and devil's claw.

This ancient way of farming is, above all, durable. Within the terms of its land and climate and of the Papago culture, it has no foreseeable end. It is an agriculture extremely conservative of its own means and possibilities. It preserves and increases the land's productivity. Fertility is built up locally, not imported. Only the annual *surplus* of water is used. There is little or no salination—an extremely serious problem in fields irrigated with groundwater. Pests and diseases are kept in check by the aridity of the climate, by the wide dispersal of fields within the region and of plants within the fields. The Papago, moreover, use many of the weeds that grow in their fields as food or fiber or medicine, and so are more apt to tolerate or encourage them than to root them out. The cycles of agriculture are in harmony with the natural cycles. The life of agriculture is renewed annually with the natural life around it, and even tends to enhance that life. There is good evidence that the traditional Indian agricultures of the Southwest increase, rather than diminish, the biological productivity and the diversity of plant and animal species. (See Amadeo M. Rea, "The Ecology of Pima Fields," *Environment Southwest,* Winter 1979; and Gary Nabhan, "The Ecology of Floodwater Farming in Arid Southwestern North America," *Agro-Ecosystems 5,* 1979. Both articles reveal what can only be called the elegance of these traditional agricultural systems.) Gary Nabhan lists nineteen domesticated plants grown in the Southwestern Indian floodwater fields, as well as thirty-three "encouraged or protected" wild plants found in the same fields.

That night we planned to sleep near the foot of Baboquivari Peak, the sacred mountain of the Papago. To get there required a long drive, part of it over a road that was only a pair of wheel tracks; we arrived late, well after dark—and not quite where we intended. But even at night, even though we were tired and a little lost, it was a place worth getting to. The moon gave enough light to make us aware of the desert sloping and widening away from us on one side, and several hills rising to straight, blunt juts of rock above us. We gathered wood for a fire, cooked supper, and ate. To me, far as I was from home and my daily concerns, the evening seemed poignantly simplified and rare. The darkness and the great quiet distances around us isolated and sharpened the fundamental feelings of weariness, hunger, and companionship.

I spread my bedroll out well away from the fire, and lay down. Now I was alone with my weariness, and the night and the silence were complete around me. It was impossible, as I lay waiting for sleep, not to feel the artificiality of my presence there. I had not arrived by any knowledge of the country, and would not survive there by any skill. That I was there was purely the result of the alien economy and technology that had destroyed the culture that belonged there. Nevertheless, I was grateful to have come and to be where I was. The night had a fine clarity. The sky held not the faintest reflection of any earthly light. There was no sound of an engine of any kind. No airplane crossed the sky all night. Off in the distance I could hear a single horned owl calling. Even in that awesome silence his voice sounded confident and large.

April 8, 1978
After breakfast we climbed up through a steep canyon, the sparse desert plants giving way to an overgrowth of brushy trees as we approached the bed of a little seasonal stream that threaded its way through the cleft. Here Karen and Gary had found a rare wild chili pepper; today they found it again, and were sorry to see that the plants had been damaged by frost.

We climbed on up then to where the bare rock rose out of the slope, following an almost hidden path that led into the mouth of a cave. We were high enough now that, looking back, we could see the broad, bright flat of the desert spreading away below us. And ahead of us the path disappeared into the cave that, against the cloudless shine of the morning light, held an almost palpable darkness. Going in required a change of sight. My companions had been there before, but they had not told me what to expect, and the full power of what I saw, once my eyes had accepted the darkness, was vouchsafed by my surprise.

The cave, I saw, was a shrine. It held only a crucifix flanked by two rather heavy wooden candelabra, and two kneeling benches in recesses on either side. The crucifix and candelabra stood on fluted pedestals sawed from saguaro cactus trunks. The body of Christ was borne on the stem of a cholla cactus whose branches had grown in a perfect cross.

I was more moved than I would have expected. I know how severely Christianity has been imposed, in the Americas and elsewhere, by conquering Europeans. The missionaries have sometimes been almost as ruthless as the soldiers, and sometimes as destructive of native cultures and possibilities. Father Eusebio Francisco Kino, the Jesuit priest who brought Christianity, Old World food plants, cattle, and horses to the Sonoran peoples between 1687 and 1711, was only one of many of his kind who could make no distinction between Christianity and colonial economics. His interest was to convert the Indians, but also to put them to work in the service of an economy that diminished both them and their land. His work, says Bowden, resulted in "two hundred years of turmoil" and "an era of constant warfare."

This shrine was the work of a priest of another sort, Father Camillus Cavignaro, who lived among the Papago in recent times; and it represented a gentler—and a truer—sort of Christianity. Here was the crucifix, not decked in the trappings of power, but tucked away in the darkness of the land itself, planted like a seed, to bear what fruit it can or will. The cave was full of the clear, pealing songs of canyon wrens feeding busily among the rocks outside. Beyond the opening, white-throated swifts turned and glided and chattered in their feeding flight, laying their intricate script on the air. And framed by the dark cave mouth the silent, brilliant day stood on the desert. It all belonged together: the dark and the light, inner and outer, joined; the body of Christ hung in sacrifice in its open sepulcher, in its intended humble union with the earth and with other bodies.

The difference is the difference between imposition and adaptation. Perhaps any innovation must be in some way disruptive. But Father Kino brought the seeds of wheat and peas, which entered quietly and, I assume, beneficently into the agriculture and the diet of the Papago. And he brought cattle, which destroyed the range and competed with the people for food. The difference, critical to our history, has received too little attention.

Leaving the cave, we saw high up in a notch on a perpendicular face of yellow rock a tiny perfect garden containing green plants, a small tree, and a patch of intensely scarlet flowers—paintbrush, we thought—as beckoning, and almost as inaccessible, as Eden.

Toward the end of the morning, we visited another neat homestead, this one tended by a quiet old man, whom we found at work in his dooryard garden. The garden (as I remember) was watered from a well. But the field we visited here had irrigation water available from a small reservoir known as a *charco*, located in such a way as to impound the surface runoff, and hold it for use during the dry season. When the water is needed, a narrow breach is dug through the dam, and the water conveyed to the crop through ditches.

From there we went to the town of Sells, where we met a young Papago farmer, Francisco Valenzuela, known as Pancho, who joined us for a trip to his fields, which lay just south of the Mexican border. Pancho is a bright, articulate man, thirty-four years old, who has returned to farming his father's fields in defiance of the drift away from tradition. He has been involved for some time in what I gathered has been a complicated and frustrating effort to persuade the Mexican government to recognize hereditary Papago rights to certain lands south of the border. The hereditary country of the Papago is divided by the international boundary, which the Papago consider a rather fanciful bit of geometry.

Pancho's effort to recover his people's land has sharpened his sense of regional and racial identity: "Just because I live north of the border and speak English or south of the border and speak Spanish"—he *does* live both places, and he speaks both languages, in addition to his own—"that doesn't mean I am an American or a Mexican. I am a Papago."

It is not common—and, I believe, not easy—for so young a man in these times to turn back to so old and difficult a discipline as traditional Papago agriculture. Pancho is a man with two minds: one, a flashy invention known as "the consumer mentality," which he shares with all modern Americans; the other, the traditional pattern of knowledge, ceremony, and community, which was the way of his ancestors for a thousand years. He is a modern man with a thousand-year-old mind. In him that old mind has begun to break its way through the new, like a corn seedling growing through a crust. And so at the time for planting the winter crops, he borrowed a tractor and drove it south to stir again the "good earth" of his ancestral fields.

The international boundary is a barbed wire fence, certainly more geometrical than geographical. The country is the same on both sides. If you didn't know that it was an international boundary, you would think it an ordinary fence. We opened an ordinary gap and drove into Mexico—wetbacks in reverse. It was pleasant to think that we had just committed an international incident, and two governments had failed to notice.

Pancho's fields were the largest that we had seen. Originally, an area of perhaps twelve acres had been cropped. So far, Pancho has brought only three or four acres back into cultivation. These now stood in wheat. The crop was headed out; in another month or so it would be cut with sickles and the grain tramped out by horses. The crop would not bear comparison with the better crops one would find growing in a well-watered region, but the quality was good, and the yield would be respectable.

Where a mesquite tree grew in the field I noticed that the wheat was greener and of somewhat better quality than in other places. The wheat undoubtedly benefits from nitrogen fixed by the roots of the mesquite, and perhaps from its shade; in addition, it has been shown that the mesquite's deep roots bring water and nutrients up to where the wheat roots can reach them. Though the mesquite was once a staple food plant of the Papago, I got the impression that they have now learned to regard it as a weed. But what I saw there made me wonder if mesquite and wheat, and perhaps other annual crops as well, should not be grown together in "two story" plantings—the annual plants benefitting from the nitrogen-fixing and other growth-enhancing abilities of the mesquite, and the mesquite from the cultivation and irrigation of the annuals.

These fields were irrigated from a *charco* much larger and deeper than the one we had looked at in the morning. And Pancho showed us another large old one, farther away, that he would like to renovate and use. He led us along scarcely legible water courses over the flat ground, methodically filling out for us his vision of a renewed agriculture in this place, based on traditional practices, native or adapted crop varieties, and surface water. I am always delighted by the patterns that a loving imagination makes as it goes to work on neglected land, and as I listened I caught his excitement. But I didn't fully understand his meaning until the next day when Gary showed me fields ruined by an agriculture based on industrial principles, exotic crops, fossil fuel, and fossil water pumped from drilled wells.

After we had walked and looked, we ate lunch in the shade of an elderberry "bush" with a trunk a foot thick, sitting with our feet in the dry ditch made long ago to carry water from the *charco* to the fields.

That was my last look at the surviving remains of traditional Papago farming. I came away with the feeling that the Papago might gain much by keeping it alive and practicing it, not as a cultural keepsake, but as a source of food and a live connection with their land and their past. But to do this, they need three things:

They need to preserve the ancient skills and knowledge, which now, like old knowledge and skills everywhere in the country, are dying with the grandparents' generation.

They need—even to preserve what the old people remember—the interest of the young people. How they will get it is not a happy question. And I think it is not a different question for the Papago than for the rest of us.

And they need suitable tools. The older farmers still use horse teams to do their fieldwork. The younger ones, I believe, at least know how to ride; saddle horses are still among the necessities here. Given this knowledge, and given the necessarily small scale of this kind of farming, the use of horses obviously makes better sense than the use of tractors. But supplies of the necessary equipment are apparently not easily available. Nor are the necessary blacksmiths, harness makers, etc. What is needed is a sort of subeconomy, modeled perhaps on that of the Amish: a local agriculture supported by the necessary local crafts and trades. I got the impression too that there would be a need for a few ranchers who would keep draft stallions or jacks to cross on the available saddle-type mares.

April 9, 1979

After a night's rest in Tucson, Gary and I started north to visit the Hopi villages. Between Tucson and Phoenix, we drove through a farm country of large, flat, irrigated fields, then being prepared for spring planting. The farmers there grow cotton and alfalfa, which are the main crops, and also pecans, sorghum, millet, sugar beets, lettuce, melons, and barley. This is modern industrial farming in its purest form: enormous, costly fields, dependent for their productivity on large machines, fossil fuels, chemical fertilizers, insecticides, and herbicides. Precarious as these dependencies are anywhere, in Arizona (as in the Southwest generally) another even more critical dependency is added: fossil water. These fields are irrigated with water pumped from ancient stores of water held in aquifers deep underground. And the aquifers are being depleted at a rate far greater than the rains can replenish them. They are being pumped down, for instance, at a rate of three feet to four feet per year at Tucson, ten feet per year at Phoenix, twenty feet per year at Beardsley. There is an annual overdraft in Arizona of 2.5 to 5 million acre-feet per year. This enormous deficit has three causes:

1. The irrigation water is used inefficiently. About half of it is lost to evaporation or wasted in other ways in transport and in the fields. (In Israel, by comparison, only about 10 percent is wasted.)

2. The agriculture is not suited or adapted to the region. In its plant varieties, methods of cultivation, etc., it is an agriculture imported intact from the well-watered eastern states—where its future is uncertain enough. It is

workable here solely because, until now, the ground-water has been abundant and fairly cheap to pump.

3. These industrial farmlands must compete for the limited water supply with a large modern urban population which, like the agriculture, has made no effort at all to adapt to the climate. Tucson and Phoenix are large cities on the eastern model, set down in the midst of the desert, enabled to survive there so far simply by the availability of subsidies of "cheap" fossil fuel and water. The people of these cities swim in swimming pools and water their lawns just like people in Cincinnati or Atlanta. Each person in Tucson uses between 125 and 175 gallons of water per day. Phoenix boasts a huge swimming pool with a "surf" of machine-made waves; it has outdoor ice skating and a professional hockey team.

To set this squandering, urban-industrial "agribusiness" against the elegantly conservative traditional agriculture of the Papago is again to illustrate the difference between imposition and adaptation—between bigotry and force on the one hand and grace and skill on the other. The Papago adapted farming to their country, and by that adaptation gave it—and themselves—the power to endure as long as the bonds between them and their land remained unbroken. The modern industrial farmers, on the other hand, have forced the country to conform to their way of farming. So long as their technology and their surplus capital can provide shortcuts, such as pumping groundwater and transporting fuels, this way will "work." But it cannot work any other way, and the signs of its failure are readily apparent to anyone who will trouble to look.

The folly of this agriculture is most plainly evident in the fields that it has already been compelled to leave behind. The reasons for this abandonment are salination, caused by the rapid evaporation of the mineral-laden groundwater, and the cost of bringing groundwater to the surface, which increases prohibitively as the aquifers are pumped lower and lower. For example, the *Arizona Daily Star* reported on July 26, 1978, that "about 40 percent of the farmland in Cochise County isn't producing this year" because many farmers could no longer afford to run their pumps. The energy costs for pumping groundwater on some farms have risen "by as much as 1,000 percent in five years." And the costs may double again by 1985.

If the abandoned fields were merely abandoned—that is, if the desert vegetation could return and cover them again—one could look at them with a much easier mind. In fact, however, abandonment is only the second stage of their degradation. The trouble is that the natural vegetation does not return to these fields. Or it does not, and cannot, do so quickly enough. For that, Gary said, desert scientists have found several reasons:

1. Salination, which prevents crops from growing, also prevents the wild plants from growing.

2. The reservoir of wild seed in the soil, which normally would start the return of native plants, has been destroyed or drastically depleted by herbicides.

3. The fields are so large that wild seeds blow in only around the edges.

4. Continuous cropping has compacted the soil and in other ways altered its natural structure.

5. The aridity of the climate would make revegetation a slow process under the best of conditions.

After the fields are abandoned, they produce only a very sparse growth of such plants as tumbleweed, cranesbill, and mustard, which cover the ground poorly. Nevertheless, the fields are then fenced and sheep are turned in to exploit their small remaining value as pasture. They are overgrazed, further exposing the ground to the winds, and allowing the dust to blow. Blowing dust has become a major traffic hazard in these areas. As the dust blows away, the heavier particles of sand and gravel stay in place. The rain beats these into a tight seal over the surface of the ground. This is the final product of "agriculture" here. It is called "desert pavement," and it is aptly named. It is as sterile as a concrete road, and feels the same underfoot.

Such fields have two possible fates: they either lie waste, at the mercy of geologic time; or they become the sites of housing developments for the retired people and vacationers who love Arizona because it is sunny and warm—so long, of course, as they can live in air-conditioned houses. The process that destroys the productivity of the land thus increases the number of mouths that must be fed. Standing in one of those miserable fields, I felt much closer to hunger than I had been at the "Hunger Conference."

One would like to say that these Arizona fields are somehow unusual, or different from the rest of American agriculture. But this *is* American agriculture, which is now fairly uniform in technology, economics, and attitude from coast to coast. This land just happens to be marginal because of its low rainfall—and it is at the margins that the weaknesses of an enterprise will show first and most dramatically. But industrial agriculture has the same precarious dependencies and is ruining land wherever it is practiced.

Because this Arizona farmland is marginal, it provides an indispensable standard by which to measure the performance of industrial agriculture. We must look at the producing fields not just in the light of their annual production, but in light of the sterile, abandoned fields lying next to them, and in light of the little Papago fields that in many centuries of use have never become sterile.

The inevitable question is: How can one presume, in the face of present population numbers, to speak so admiringly of an agriculture that, at best, sustained a population of only a few thousands? One can do so, first, because the traditional Papago agriculture is *exemplary*. It provides us a far better example or pattern and a better set of standards for agriculture than does the factory, which has provided the pattern of the "agribusiness" farm. Second, one can do so because the Papago fields will prove more productive in the long run than the "agribusiness" fields, even though their annual per acre productivity is smaller. The critical point is that Papago agriculture incorporates in itself an entirely competent understanding that Papago land is agriculturally marginal, whereas "agribusiness" on even more marginal land incorporates an elaborate pretense that desert land is *not* marginal. It is better to sustain a small population indefinitely than to build up a large artificial population on an agricultural system of which the basic principle is its willingness to destroy itself.

The squandering of Arizona's land and water is typical behavior with us. It is being done on the altogether conventional assumption—the nearest we come, now, to a national faith—that before the invited disaster can complete itself some technological "breakthrough" will produce a remedy. Charles Bowden's book contains a statement by Aaron Wiener (published in 1972) which perfectly expresses this assumption. The exhaustion of the stores of groundwater, Wiener said, "will create new economic assets that might make it economically feasible either to replace in the future the then-exhausted supply by the import of water (natural or man-made) or, alternatively, to adopt a change in the resources base that would substitute other inputs for water or at least significantly reduce water requirements." The desperation of this statement is only very thinly concealed by its stupidity. By this same sort of reasoning, one could justify gambling away one's farm on the ground that an as-yet-unknown rich uncle might soon die and leave one a fortune in bank stock.

Proposed "breakthroughs" so far include importing

water from rivers in the eastern United States or Canada and towing antarctic icebergs into the Gulf of California. This too is altogether conventional: an industrial problem in one place may properly be solved by creating problems in a different place. Of course, the farther from home you go to solve your problem, the more expensive the solution will be. The cost of continued water affluence in the arid Southwest will probably be economic peonage for its captive urban population.

In our long drive northward we again moved through the marks of the depletion and ruin of a new society to the conserved but threatened remnants of an old one. We ate lunch at Tuba City, just to the west of the Hopi reservation, and went first to the fields belonging to the village of Lower Moenkopi. These, unlike most Hopi fields, are irrigated from a reservoir and from springs. They step in terraces down the tilted floor of a canyon, fanning out below onto the bottomlands along a small river. Feathered prayer sticks stood in the fields, which were almost deserted, the spring crops not yet planted.

We left the car and, seeing one man at work beside an irrigation ditch, walked over to talk to him. This farmer had just finished watering a newly planted small patch of onions. The day was overcast, windy, and cool, and we were all wearing our jackets. We introduced ourselves.

"Have you come to experiment on the Hopi?"

"No."

"Why not? Everybody else does."

This was said with a laugh, but it was a measure of defense, perhaps of warning. Its effect was to make a precise, essential distinction: he belonged there; we did not. I liked him for this, and found it maybe as useful as he did; he had given us permission to be strangers.

He was a man perhaps in his sixties. Unlike the elderly

Papago men we had met, he was a ready conversationalist. He was a member of the tribal council, and had been to Washington on his people's business. He enjoyed telling me that he had flown over my state of Kentucky.

Once he was assured that we weren't "experimenters" and weren't sniffing for tribal secrets, but only wanted to talk about farming and look at the fields, the encounter became a rather usual kind of visit.

We asked how soon the planting would be done.

Not for a while, the councilman said. They needed warmer weather. The weather was disturbingly cold for the time of year. He said that his people's ancestors had prophesied that they would one day plant their crops wearing coats and gloves. That this now appeared to be coming true he obviously found both disturbing and satisfying. But he also made several rather mundane jokes, more familiar to us than to his ancestors, and talked with us about the fields, the crops, and the irrigation system. At the end of our visit Gary gave him some seeds, and promised others. Like Steve Brush in Peru, Gary Nabhan is a seed carrier and a seed exchanger, distributing varieties and species from place to place as the medieval minstrels distributed songs. Thus the stranger assumes a cultural role by having something to offer. Sometimes Gary has the pleasure of bringing the seeds of a traditional crop back to where it was once grown but has been lost.

Like all small fields that are intensively used and intensively cared for, those at Lower Moenkopi were pleasing to look at. This was so even though they were unplanted. Their bareness allowed one to see how skillfully they had been shaped to conserve the land and hold the water, and it gave them a strict formal beauty. In these fields and on their borders are grown plum, pear, peach, and apple trees, grapes, corn, amaranth, sunflowers, white tepary beans, limas, and two other varieties of beans, a yellow and a purple.

Approaching Hotevilla on Third Mesa we began to see the dry farming that is more characteristic of the Hopi. In a few places besides Moenkopi they have water available for irrigation, but mainly they have had to shape their farming to the hard conditions of the desert. In Hopi country rainfall averages only nine inches a year, and drought and frost limit the cropping season to about ninety days. These circumstances require careful adaptations both of crop varieties and of farming methods.

Like the Papago farmers, the dry-land Hopi farmers locate their fields where the groundwater accumulates and is held near the surface. To assure sufficient moisture, they plant deep—eight inches or more—and space the planting hills widely apart to assure a proper balance between the plant population and the available water. It is a situation in which minimum tillage would obviously make sense, exposing the least possible amount of soil moisture to the drying sunlight, and in fact the old-time Hopi did not plow. Now some of the larger, flatter fields are plowed with tractors.

Some of the Hopi have fields as large as ten acres or more, but those that we saw from the road near Hotevilla were little hillside fields among the rocks, where undoubtedly most of the work is still done by hand. These fields too were still unplanted. In some of them were woven brush fences used as windbreaks. The crops here are corn, beans, grapes, and fruit trees. This is a windy country and the sandy soil is unstable. Some of the old peach trees had their roots exposed to a depth of three or four feet; others were buried to their branch ends, so that the sand dunes bristled with blossoming twigs.

By the time Gary completed an errand in Hotevilla, it was twilight. We drove on to Old Oraibi, which has the reputation of being the most conservative of the Hopi towns. Gary thought we might be forbidden to enter, but we turned off the road anyhow, only to encounter the following sign:

WARNING   WARNING

No outside white visitors allowed. Because of your failure to obey the laws of our tribe as well as the laws of your own this village is hereby closed.

We thought it a respectable sign, and we respected it. We turned around and drove on to the motel near New Oraibi, where we would spend the night.

The afternoon had produced several encounters with Hopi people, all of them casual, some of them lasting only a few minutes, but I was nevertheless impressed by them—as I would be again the next day. They are a people of great beauty, dignity, and courtesy. They have been for a long time the objects of study and curiosity, but I saw no one who seemed willing to indulge or invite this. Instead, it has made them careful and reserved. But with the reserve goes a courtesy so fine and considerate that, even while you are held at an appropriate distance, you are given the same respect that is required of you. Also I thought them extraordinarily well spoken; they talked quietly, without exaggeration of expression or waste of words, forming their sentences with consideration and with care.

But I was also impressed by the amount of junk scattered around the villages—the usual modern American assortment of cans and bottles, plastic jugs, old cars, etc. In that climate even paper lasts a long time, and it blows everywhere. The junk surprised me; most people who write or talk about Indians, I think, try to see or imagine them apart from the worst—or at least the most unsightly—influences of white society. But of course one should not be surprised. When junk is everywhere—better hidden in some places than in others—why should one not expect to find it here? What was wrong, I thought, with the warning sign at Old Oraibi was that it was not exclusive enough. What use is it to keep out white people if you let in their "consumer mentality" and the wasteful habits and the imperishable garbage that go with it?

As well as I could tell from a distance, however, it did seem to me that Old Oraibi was less littered than the other villages I saw.

April 10, 1979
While it was still fairly early in the morning we went to visit an elderly couple, friends of Gary, who live at the Second Mesa village of Shongopovi. We were welcomed into an orderly kitchen full of sunlight, warmed by a small wood-burning stove. There occurred an exchange of news and seeds. Our hosts had found what appeared to be a new bean variety in their field. They were excited about it, and gave Gary some of the seed.

We described the peach trees with wind-exposed roots that we had seen the evening before, and were told that there was a remedy for that: you can pile weeds around the roots to catch the drifting sand, and so cover them up again. As we listened, the old man made us aware of the complex soil husbandry of the Hopi dry-land agriculture. In the desert one may be at the mercy of some things, but not of everything. Not of the wind. The brush fences we had seen were also part of the ancient system. They were woven of sagebrush and rabbit brush to keep the blowing sand from covering the planted crop rows. The sand that caught behind them then became the crop rows of the next year.

Now, he said, "The Hopi are getting lazy." They don't do so much of that sort of work any more; don't take care of such things as exposed peach tree roots.

The old man still plants his fields every year. "I enjoy being out there," he said. His tools are a hoe and a push plow. But many of the younger people don't farm any more. They work at jobs instead. In the farming that is done there is increasing use of tractors with increasing erosion during the months when the soil is most vulnerable. They no longer use horses. They no longer have a blacksmith.

When the old man went to the Indian school in 1915, a farmer was there who taught useful, practical things, "and not from books."

"What do they teach now?"

"Nothing, I think."

The young, at any rate, are not learning the ancient skills of survival, but are growing up under the influence of the dominant society. By the middle of this century, Edward H. Spicer wrote in *Cycles of Conquest,* the Pueblo were "an enclave of farming people within the United States economy, comparable to such enclaves as the Old Order Amish ... with the important difference that they did not have their own schools." The drift is clear. The modern schools may or may not teach "nothing," but—among the Hopi, as among the rest of us—they do act as powerful agents of "the United States economy." They do not prepare the young people to stay at home and make the most of the best local possibilities. They serve the idea that it is good to produce little and consume much.

And so now, the old man said, the young don't go to the fields any more, because the fields are below the mesa, and one gets there by footpath. The young don't go where they can't drive. They are thus forsaking their inheritance of knowledge and skill and "will have to learn the hard way."

Except for its humor and lack of bitterness, the old man's talk was similar to talk I have heard from many old people: the times are changing for the worse; the young are forsaking the old ways; hard times will come. There is a tendency to discredit such talk "because old men have always said that." But I think that in our time it cannot be so lightly dismissed. It can, of course, be a kind of mindless rant. But this old man was not ranting. He was talking about what he had observed of the disintegration of a local culture and the decline of values.

What one thinks of this disintegration and decline will depend on one's opinion of the United States economy, and on one's confidence in it. If one believes that it is better to buy food than to grow it, then one is not going to worry about the decline of any particular farming community, especially if that community is based on subsistence farming.

I agree with the old man. I *am* worried about the decline of farming communities of all kinds, because I think that among the practical consequences of that decline will sooner or later be hunger.

In some respects, the traditional subsistence agricultures are the best agricultures, the best assurances of a continuous food supply, simply because they are not—or were not—dependent on outside sources that must be purchased. To exchange these locally self-sufficient subsistence agricultures for the "good life" of a consumer economy is like climbing out of a lifeboat onto a sinking ship. That image, I think, only *seems* extravagant. The values of our present economy do indeed suggest that it is better to perish with some ostentation of fashion and expense than to survive by modest competence, thrift, and industry.

In saying such things, one must anticipate the accusation that one is simply indulging in nostalgia—sentimentalizing the past, yearning naively for the survival of quaint anachronisms and relics. That might be true if one were dealing only with rare and isolated instances. The fact is, however, that these instances are not rare or isolated. The decline of the Indian agricultures of the Southwest follows exactly the pattern of the decline of local agricultures everywhere else in the country. The economy of extravagance has over-thrown the economies of thrift. Local cultures and agricul-tures such as those of the Hopi and the Papago do not de-serve to survive for their picturesque trappings or their in-terest as artifacts; they deserve to survive—and to be em-ulated—because they embody the principles of thrift and care that are indispensable to the survival of human beings.

Our last stop on the Hopi reservation was at Walpi, at the southern tip of First Mesa. This village is built on a small, high table of rock, surrounded by steep cliffs. It was moved here in fear of Spanish reprisal after the Pueblo Revolt of 1680. The place is a natural fortress, eminently defensible. With its stone houses and narrow, crooked streets, it has something of the feeling of the fortified hill towns of south-ern France.

It is a place of considerable beauty in itself, having the propriety and grace that come with respect for a building site. And it is surrounded by immense breadths of desert country at once lovely and austere. But around the base of the cliffs, surrounding the village, cutting it off, at least vis-ually, from its setting, is a sort of belt line of castover junk: cans and bottles, paper, plastic, old bedsprings, etc. It looks as though Walpi has indeed just survived a siege, in which the opposing armies hurled refuse at each other.

It was a cold day, and the air was full of Peabody coal smoke.

On our way back to Tucson, driving down a narrow river valley in the northeast corner of the Sonora Desert, we saw one more human habitation that seemed to us to belong where it was. This was a set of ranch buildings against the foot of a steep hill on the far side of the river. On the bottomland between the buildings and the river were alfalfa fields irrigated with diverted river water. These furnished winter pasture to—we estimated—three or four hundred head of Santa Gertrudis and Hereford cattle. In summer the cattle would go up onto the hilly rangeland, leaving the alfalfa fields to be harvested for hay. The fields were green, the buildings all painted and well kept.

It was getting late, and we had no time to cross the river and look around and talk to the people who live there. But we did stop on the roadside and spend a while looking at the place and taking pleasure in it. The Indian fields we had looked at represented two proven ways that allowed permanent human use of the land. This ranch looked to us as though it might be another. It was lovely, and more lovely for being rare.

3

The Native Grasses
and What They Mean
(1979)

On a bright weekend in mid-October I put myself into the care of Timothy Taylor, agronomist and authority on grasslands at the University of Kentucky, and Bill Martin, partisan of plant communities and ecologist of Eastern Kentucky University. Our project was a tour of the fugitive survivals and remnants of the native tallgrass prairie that, before the invasion of white people, flourished in the so-called "barrens" of western Kentucky, in the meadows of the region now called the "Bluegrass," and in scattered clearings and savannahs on, probably, to the eastern seaboard.

In three days we drove some seven hundred miles. West of Elizabethtown we made a long loop that carried us as far as the town of Benton in Marshall County. In general we followed the Western Kentucky Parkway westward, and returned eastward along Highway 68 and I-65. But we made many side trips, mainly to visit the tiny and usually improbable enclaves where the members of the once-dominant plant community of the tallgrass prairie have taken refuge and survived, as Bill Martin says, "by the skin of their teeth." These places have been lovingly sought, studied, brooded over, and worried about by such people

77

as Bill and Tim and the western Kentucky botanist Raymond Athey.

Late Friday morning we waded through a narrow strip of "railroad prairie" along the Illinois Central tracks not far west of Elizabethtown. On Saturday we walked for hours in abandoned fields in TVA's Land Between the Lakes, where the prairie plants are making a straggling comeback in competition with persimmon, sassafras, redbud and other "pioneer" trees. But in Land Between the Lakes these areas are being managed to favor the prairie plants by Lawrence Philpot, an able, likable agronomist who oversees all agricultural practices in this huge "recreation area." In consultation with Tim and Bill, he has burned these fields several times as a way of holding the trees in check. The three of them spent time that Saturday planning for the next burn, and also for a harvest of grass seed for Tim's experimental plots at the University.

For a while that morning Raymond Athey accompanied us. He is a slender man, quietly companionable and alert. He is in the carpet business at Paducah, but has, on his own, made himself a botanist and plant taxonomist, and is now probably the chief authority on the plants of the region. He is, even more downrightly than Bill Martin, a partisan of the community of the prairie plants; in the areas that were originally grasslands, he says, "the trees are the enemy." He had a botanical errand more urgent than ours, and left us before noon. But before he went he took us along a dry creek bed to a place where muscadine grapes hung ripe from the low trees.

Just before noon, we all climbed into the back of Lawrence Philpot's pickup truck, and he drove us out for a look at the buffalo herd that is confined on an ample acreage of grassland behind a six-foot wire fence. We respectfully kept our distance, and the buffalo kept theirs. But the distance was not great. These are "wild animals" by definition, and they *are* more problematic than cattle—when hauling them, for instance, it is necessary to keep the truck moving, for in tight confinement without the distraction of motion, they will fight and injure or kill each other. But they are not *very* wild. They ambled around, eyeing us curiously, within a few feet of the truck. The night before, at the tourist lodgings at Brandon Spring, we had eaten patties of ground buffalo meat, and thought it easily as good as beef.

These are "woodland" buffalo, but they are not browsers, and of course never were. They are grazers, and their presence originally in the eastern parts of the country argues the availability of grass, of meadows, even where the land was predominantly forested. And where they grazed, at this latitude, they probably grazed the tall prairie grasses. At Land Between the Lakes they were grazing on fescue. They are impressive animals, and one looks at them necessarily with an awakened sense of history. But for one who hankers as I do for a glimpse of the country as it was in, say, 1750, a good deal of imagining is still necessary. You have to imagine the animals to begin with, in much larger numbers—in hundreds and thousands instead of forties and fifties. You have to imagine the fences gone, the trees gone, the prospects lengthened. And you have to imagine the buffalo feeding, not on the ankle-high fall pasture of fescue, but in grasses just maturing to seed, heading out higher than the withers of the biggest bull.

Belonging in the prairie plant community is a considerable variety of broad-leafed plants. Some of these are spectacular wildflowers, but in Kentucky in October the season of bloom is nearly past. We saw blazing star, several varieties of goldenrod and aster, a lovely small lobelia. There are also shrubs of various kinds, and dwarf trees such as crab apple and hackberry. But the overwhelming visual impression is made by the grass. In the places we visited, we invariably saw the little and big bluestems and Indian grass. At this season they are turning brown, but it is the liveliest, warmest of browns, with a strong glow of red in it. To walk in a stand of big bluestem or Indian grass is to be submerged. It is an enclosing element, different in feeling both from forest and from conventional pasture land. If it were necessary to go any considerable distance in it, it would be good to be on horseback.

Of these species, Indian grass, though beautiful, is the least palatable. The bluestems are attractive to grazers and are nutritious, good both for pasture and for "prairie hay." Another tall grass once probably native to the region, but which we saw nowhere on our journey, is switchgrass, a species Tim thinks highly promising for summer pasture, and which he has been growing extensively in test plots on the University of Kentucky's Spindletop Farm at Lexington. But his seed has come from West Virginia, not Kentucky.

Land Between the Lakes is a tract of 170,000 acres, lying between Kentucky Lake and Lake Barkley on the Tennessee and Cumberland Rivers. TVA took it over about a decade ago as a "demonstration in conservation-based recreation." Of the 170,000 acres, 75,000 were already publicly owned. The rest—95,000 acres—was comprised of farms, homesites, and timber lands belonging to, among others, 949 "resident families"—2,738 people in all. These people were moved out because, according to a TVA ruling, private holdings "would be a deterrent to maximum public use of the area." They were, of course, "compensated," but for the most part, they were strongly attached to their homes and communities, and gave them up in grief and in protest. Those who would not consent to the price offered by TVA were forced out by condemnation under the "right of eminent domain." The removal of these families was justified by one TVA official partly on the ground that their way of life "never quite succeeded."

All its bureaucratic solemnity about population growth, open space, recreation, and conservation would be more plausible if TVA had not been, at the time, perhaps the largest of the users of strip-mined coal, and was thus assenting to and abetting the destruction of far more land and "open space" than it was conserving, some of it much nearer to the centers of population than Land Between the Lakes.

It is an ugly story of the tyranny of "public service"—the homes of "the few" high-mindedly sacrificed to the "recreation" of "the many." Once this obliteration of the settled human life of the place has been forgotten, Bill Martin says, then it may be possible to be simply grateful for this large nature preserve. But he says so, knowing what will be lost in that forgetting. Pleased as I was to see the buffalo and the woods and the renewing meadows of tall grass, I would much have preferred to see the 2,738 people back at home. But we have no bureau for accomplishing that.

On Sunday morning we visited three more prairie survival sites, widely scattered along our homeward route. One, the largest, was an abandoned farm where the tall grasses have reclaimed the droughty, thin-soiled fields, and were holding on among scattered cedars and scrubby oaks, buckthorns, dwarf hackberries, and gnarled, lichened little crab apples like bonsai.

Another was a tiny rocky hillside patch almost overgrown by cedars on a sideroad near Mammoth Cave. Here Bill Martin began a concentrated search low to the ground, and after a while we heard his voice full of a discoverer's elation: "There it is! There it is!" He had found a few tufts of the prairie grass known as side-oats grama, the only example we saw.

The most moving and memorable of these plots was a stony roadside that we came to in early morning while it was still in shadow and moist with dew. This was too infertile a place to support a thick stand of the grasses, but, perhaps for that reason, it had the richest variety of broadleafed plants. It has survived by the care of a farm family whose modest house stands a short way up the road. They mow this roadside every year to keep the highway department from spraying it. The wife thinks the flowers are beautiful, and she once told Bill, "I don't know what they are, but I don't want anybody fooling with them."

What is the importance or value of these perhaps ephemeral relics of a long-vanished prairie? What do they signify? In a sense, I think, they serve us as "control plots"—as do, in other areas, the equally rare and threatened stands of virgin forest. The condition of the land as it was when we came to it is the only possible measure of our history. Only by knowing what it was can we tell to what point or result it has been changed.

As we felled and burned the forests, so we burned, plowed, and overgrazed the prairies. We came with visions, but not with sight. We did not see or understand where we were or what was there, but destroyed what was there for the sake of what we desired. And the desire was always native to the place we had left behind.

The forest could not survive because we did not see it; we saw cleared fields. The prairies could not survive because in their place we saw cornfields and pastures sowed to the cool-season grasses of the Old World. And this habit of assigning a higher value to what might be than to what is has stayed with us, so that we have continued to sacrifice the health of our land and of our own communities to the abstract values of money making and industrialism. Or to "recreation"; it is because we have so insulted and despoiled Creation that we need recreation. In the last generation huge areas have been laid waste by strip mining. And it is no mere coincidence that the spread of surface mining has been paralleled by the spread of extractive agriculture.

To see and respect what is there is the first duty of stewardship. "I don't know what they are," the farm wife said to Bill Martin, "but I don't want anybody fooling with them." That is an ecological principle, and a religious one. If you don't know what it is, don't fool with it. Don't use it carelessly. Don't destroy it. And who knows in any ultimate or final sense what any creature is? The biochemist Erwin Chargaff has written that "Even the most exact of our exact sciences float above . . . abysses that cannot be explored."

When we destroyed the native prairie, what did we destroy? Was it merely a curiosity, a "natural wonder" of some sort? Hardly. Tim Taylor's work very forcibly suggests the practical significance of the loss. The grasses of the tallgrass prairie, to begin with, are warm-season grasses. That is, they make their most vigorous growth during the hot summer months when the cool-season grasses such as fescue and bluegrass are semi-dormant. And the notorious weakness of our pasture economy, as it now stands, is that we have few grasses of high quality in current use that grow well in the hot months.

The prairie grasses, moreover, are extremely efficient users of light—almost twice as efficient as, say, fescue. This means that their productivity—of pasture, hay, or humus—is spectacularly greater than that of the cool-season grasses. On Tim's test plots switchgrass without added nitrogen produced 5,600 pounds of dry matter per acre; with 69 pounds of supplemental nitrogen per acre, the production was increased to 9,600 pounds. Tall fescue, by comparison, produced 2,400 pounds without nitrogen, and with 100 pounds of added nitrogen per acre, it yielded 5,400 pounds. Yearling steers tested in North Dakota gained two pounds per day on switchgrass, and almost as much on bluestem.

And so these grasses may be said to be highly promising sources of pasture and hay. They are, in addition, excellent soil builders, and they provide excellent cover. But to use them properly, to preserve the stands in use, and to integrate them satisfactorily into farm pasture programs, Tim says, will require highly skillful and careful management.

And how capable is our agriculture, in its present state, of highly skillful and careful management? Not very, I am afraid. What I saw on this trip through western Kentucky confirms what I have been constrained to conclude from agricultural travels in many other states: that this country is now poorly farmed, the land used less skillfully and carefully than ever before. In general, the better the land, the neater the farms look, but this is the monotonous, sterile neatness of monoculture. The corn rows are long and straight, oblivious of the contour of the land and of the paths of drainage. Often the fences are gone, which means that the livestock is gone, which means that to produce an income the fields must now be continuously cropped.

The rougher the land, the worse it is neglected and the harder it is used. We saw many usable pastures gone to bushes or overgrazed. We took a close look at a soybean field on rolling land that ought to have been in permanent pasture. The soybean plant is hard on sloping ground because it loosens the soil and makes it easy to wash. In this field the rows ran straight downhill to the waterways, which had been plowed and planted like the rest. There were washes up to six inches deep between the rows. The waterways were wide gulleys twelve to sixteen inches deep. This sort of thing may be attributed to the use of large, high-speed equipment. But it has an antecedent cause: a mind willing to accept permanent loss as a tolerable charge against annual gain.

We saw many bean fields so weedy as to be nearly unidentifiable, and many others cut to pieces by heavy harvest equipment passing over them when they were wet. And we saw, to my mind, far too much land in corn, some of it too steep for cultivation, most of it obviously not to be harvested in time to be protected over the winter by a cover crop.

In the whole trip we saw not a single sheep. Not one. Kentucky, a state well suited to sheep, and once abundantly productive of them, now has only 24,000 head—insignificant by comparison even with its present population of 70,000 deer.

As we drove through these deteriorating farmlands on Sunday morning, Tim described for us his father's way of managing his cornfields. In the fields nearest the house, in addition to corn, he planted beans and pumpkins. He sowed cowpeas between the corn hills to improve the soil. Before picking the corn, he sowed redtop to provide a winter cover. He harvested the corn in three stages. First, he stripped off the blades below the ear; this was called "blade fodder." Next, he cut off the stalk above the ear, which was "top fodder." Both kinds of fodder were carefully stored for winter feed. Finally he harvested the ears. After the harvest, the cattle were turned in to glean whatever grain or forage may have been left. On a frozen morning the next spring, the field was sowed to red clover.

This was Tim's childhood experience of good farming,

and it left its mark on him. It was a native farming method at once skillful, thrifty, respectful of the land, and saving of it. To describe its complexity and intelligence, its wakeful determination to preserve the field while using it, one can only say that it was elegant. And it has nearly vanished from the world. The "agribusiness" version of farming, by comparison, is crude. About as elegant as a transparent spittoon.

II

4

The International
Hill Land Symposium
in West Virginia
(1976)

I attended the first three days of a week-long International
Hill Land Symposium, held October 3 – 9 in Morgantown,
West Virginia. The Symposium was sponsored by West
Virginia University in cooperation with the United States
Agency for International Development; the United States
Department of Agriculture—Agricultural Research Ser-
vice; the American Forage and Grassland Council; the
West Virginia State Departments of Agriculture, Com-
merce, and Natural Resources; and the Benedum Founda-
tion.

The subject was certainly important enough to merit
such sponsorship. And if the meeting frequently lacked
enough complexity and depth, it was appropriately large in
scale. It brought together scientists from twenty-six foreign
countries and twenty-two states. The various sessions of
the Symposium were tightly scheduled from 8:15 in the
morning until past 5:00 in the afternoon to permit the de-
livery of 137 papers. There is no doubt that the Symposium
succeeded in producing a voluminous exchange of expert
information, much of it highly interesting and of great potential usefulness. And it gave one a disturbing awareness
of the immensity and difficulty of the problems of food

production on steep land. I left the meeting, however, with a considerable doubt as to the amount of good that may come of it—at least in this country. For the Symposium also had an unintended value as an exhibition of the alarming limitations and weaknesses of specialized agricultural thinking.

The sessions of the conference—most of which ran simultaneously with one or two other sessions, so that one always had to be chosen at the expense of another—were divided neatly into specialized concerns and disciplines: Hill Land Improvement Techniques, Social and Cultural Relationships Influencing Hill Land Use, Economics of Hill Land Utilization, Crop Production Systems on Hill Land, Disturbed Land Areas, Hill Ecosystems, Animal Production Systems on Hill Lands, Multiple Use of Hill Lands. The Symposium thus imposed on the problem of hill farming, not the necessary responsiveness of interdisciplinary intelligence, but the rigidly departmentalized structure of the Modern American university.

And this structure of academic incoherence was then placed in the swanky, expensive Lakeview Inn and Country Club, and further set apart by a registration fee of $40.00. (My four nights and three days there cost, in all, $194.11.)

By arrangement, then, the experts of the conference were divided not only from each other along disciplinary lines, but also from the hill farmers whose problems they were talking about. Perhaps the most significant (and ominous) aspect of the meeting was the absence of farmers. The idea of a small hill farmer attending that meeting—not only expensive, but forbidding a coatless man to be served dinner in the dining room—was simply preposterous. There were, so far as I could make out, only two or three West Virginia county agents. And the students who attended were there mainly to run the projectors and sound equipment. The experts were talking to each other in the sealed greenhouse of expert prestige, and many important questions were un-asked.

I am aware, of course, that there are good reasons to limit the attendance at such a meeting. More people might have overcrowded the rooms and made the discussions unwieldy. One can certainly see a value in keeping the dialogue as unencumbered as possible—especially in view of the international urgency of the subject. And yet I cannot escape the feeling that both the exclusiveness and the departmentalization of this meeting were a fair representation of the way our agricultural specialists think and conduct their business. Farmers were absent, it seemed to me, not just because they could not afford to be there or because they might have overcrowded the sessions or disrupted them with their questions or criticisms—they were absent because they were not being thought about.

I heard several American experts read papers on agricultural technologies and techniques, but without ever mentioning their economic effects on the farmers who used them—much less their political, social, and cultural effects. The interest was simply in what would "work" within the tight confines of the expert's specialty. We would be shown a slide of an improved hillside on a farm, say, in Ohio. We would be told the composition of the soil, the techniques of renovation, the forage plants used. But we would not be told the history of the field, or who it belonged to, or how large the farm was, or how the field looked *before* it was improved, or what it cost to improve it, or whether or not its improvement was profitable to the farmer.

One is left with the conviction that these experts talking to each other over their Roast Long Island Duck with Sauce Bigarade and Filet Mignon Bernaise were exemplifying problems greater than any they had solved: How are we to keep knowledge from being bottled up in the bureaus and universities? How are we to get the solutions within reach of the problems? How are we to test knowledge in practice, in the lives of people, as well as in the controlled conditions of laboratories and experimental farms? How do we bring disciplines together? How do we learn to subject technology to the measurement and restraint of cultural value?

It is certain that as long as expert knowledge remains in the heads of experts it cannot become a solution. This dilemma, in fact, was commented upon by several participants in the Symposium. Roy Hughes of the Welsh Plant Breeding Station said that there is a greater need for application than for research. S. B. Nepali, of Nepal, told us that in his country it is difficult to get new technology into the high mountains because the technologists do not like to go to such remote places. M. L. Barnett, of Cornell University, spoke of the need for experts to stay in the field and talk *with* (not *to*) the farmers; this seemed to be related, in his mind, to the need for collaboration, the crossing of disciplinary lines, in addition to specialization. And William C. Thiesenhusen, of the University of Wisconsin, Madison, reminded his hearers "of the futility of studying problems of humans as if they were packaged as disciplinary units," and warned "that the solution to the problem [of the preservation of hillsides in cultivation] depends as much or more on social organization and on economics than on a new technology."

"The emphasis of this conference," Professor Thiesenhusen continued, "has been on discovering new technical information to help save the hillsides. . . . But what is of prime importance . . . is using what is now known about soil conservation. And what is now known and what is yet to be discovered will not be applied if the social, political, and economic organization [is] not appropriate for the task." And, turning to the problem already raised by Mr. Nepali, Professor Thiesenhusen asked: "How many soil-conservation experts are there who are available to work with, say, 500 small farmers who crowd onto one hillside on which the minimal slope is 25 percent?"

Of the papers I heard that addressed these problems, Professor Thiesenhusen's was easily the best argued. He alone, for example, confronted the issue of land tenure: "What agricultural laborer with a year-to-year lease on a farm will go to great lengths to save hillside soil if he knows that his back-breaking efforts at terracing will merely save the resource for the landowner, who may dispossess him as soon as his laborious work is complete? What squatter will practice contour farming or reforestation if his occupancy rights are so tenuous that it is cheaper for him to move on to another hillside patch next year?"

And so beneath the division into specialized disciplines that was *supposedly* the structure of the meeting, I felt the working of another division that seemed to me to be its *real* structure. This was a division between the knowledge coming out of small, heavily populated, intensely cultivated countries all over the world and the knowledge coming out of the specialized, technological obsessions of "agribusiness."

The presentations of the agrispecialists tended to focus narrowly upon the efficacy of new techniques, varieties, or methods. And this narrowness tended to sustain an implication or a tone of optimism: the innovation in question was good or hopeful or promising insofar as, within a tidy experimental context, it could be said to "work."

The other side, by contrast, tended to be uneasily aware of the *consequences* of technological innovation. And it tended to see human life and satisfaction as sustained by a *balance* of cultural and natural forces and orders, not by a simple series of scientific "breakthroughs."

Stephen Brush, an anthropologist at the College of William and Mary, in another of the Symposium's exceptional papers, gave a description of the complex and sophisticated potato agriculture of the Andean peasants of Peru. These people still grow and preserve their crops, without modern chemicals or techniques, by an intricate system of adjustments among varieties, planting, climatic zones, etc. They fertilize their crops with guano and sheep manure (by penning sheep in the fields). They protect their land against erosion by keeping the fields small, by the use of hedgerows and horizontal plowing, and by field rotation. They protect their crops against pests and diseases, and against climatic extremities, by the greatest possible diversity of plant varieties and by diversified strategies of cultivation which include crop rotation and fallowing.

Professor Brush then spoke of the revolutionary influence upon this traditional culture of an innovative high-yield technology and "improved" varieties, which require a money economy and credit, favor big producers, and threaten to destroy both the human community and the ecological viability of a farming system that is "the result of thousands of years of natural and human selection." Although the "improved" potato varieties may yield two or three times as much as the traditional varieties, they do not taste as good, are less marketable, and are significantly less nutritious. Like Professor Barnett, Professor Brush warned that such an agricultural system cannot be made better by anyone who does not thoroughly understand and respect it as it is.

Another speaker who exhibited an insight into the complex relations between technology and society was R. R. Appadurai of the University of Sri Lanka. Sri Lanka is a small country emerging from a history of colonialism, with a rapidly growing population, and high unemployment. The country has begun an extensive program of land reform, agricultural diversification, and soil conservation. Most interesting, however, is its policy of limiting the use of technology. Mechanical technology is felt to be both unnecessary and potentially harmful, especially insofar as it would tend to put people out of work. The government's policy is to promote labor-intensive methods in order to make as full a use as possible of the available human energy. The goal is a *stable* agriculture, and so the strategy is to concentrate on the use of resources immediately available in the land and the people.

And this sort of thinking is by no means limited to the so-called "developing nations." According to the testimony of several of the speakers, the hill farms of the United Kingdom, Switzerland, and other Central European countries are maintained by subsidies at times when they are not profitable enough to survive on their own within the economy. These "marginal" farms and their farmers are looked upon as vital resources that will be needed in times of crisis, and so policies have been evolved to keep them productive. By contrast, the small farms of this country that may be "marginal" within current economic conditions are simply regarded as "inefficient" and allowed to fall into disuse. We are similarly wasting thousands of young farmers who, because of the high costs of land, interest, and equipment, cannot afford to farm.

I was impressed to see how carefully the small countries have based their agricultural thinking on the memory and the anticipation of crisis. And I was equally impressed to see how little attention was paid to that possibility by the American experts whose papers I listened to—when our agriculture, with its absolute dependence on petroleum and machine technology, is probably the most endangered of any in the world. It was clear that those agricultural systems based upon the use of human energy and local resources could survive the sort of crisis that many of them already had survived, whereas the present American system of agriculture will fail if the fuel tanks run dry.

This contrast became most vivid in the last of the sessions I attended, which was concerned with various methods of terracing hillsides. We had heard a paper from Italy and two from Taiwan. All three of these described methods using machines, but also making use of small-scale, intensive management, with intercropping, mulching, etc. We then heard a paper by Professor H. Wittmuss of the University of Nebraska, which seemed to me to exemplify our expert agricultural thinking at its worst. Professor Wittmuss was talking about a sloping landscape of deep loess soil and fertile subsoil. The idea was to terrace this land for the purpose of controlling erosion, and at the same time to make it efficiently farmable with the huge machines of contemporary grainbelt agriculture. This meant that the "nonparallel areas" between the terraces should be kept to a minimum—that is, the terraces themselves were required to be parallel, which in turn required the hillsides to be bulldozed to a uniformity of shape and contour. Professor Wittmuss described a method of terracing based upon the use of computer designs on topographic maps made by "aerial photography and stereographic plotting." This was successful in that it "reduced the soil loss as much as 95 percent compared to up-and-down hill farming and reduced the nonparallel area from 24.2 percent to 2 percent of the terraced area." Characteristically, Professor Wittmuss did not discuss the social, cultural, or political implications of this method. He did not say by whom or in what circumstances it could be afforded. The only measures of "success" were the efficiency of reducing erosion and the degree to which the landscape would be made to conform to a geometry dictated by machines.

Asked about the possibility of growing trees on the grassed backslopes of the terraces, Professor Wittmuss replied that it was impossible because the trees would compete with the row crops. When it was pointed out that the Symposium had produced an abundance of evidence of the desirability and the success of interplantings of trees and row crops, Professor Wittmuss replied that the herbicides sprayed on the row crops would kill the trees. Thus a useful and obvious economy was overruled simply because it *apparently* would not conform to a dominant technology.

Because it depends on unusually deep soil, Professor Wittmuss's method is not universally applicable—for which we may thank our stars. But it is nevertheless representative of our experts' willingness to let the solution define the problem, and to exclude from consideration all but the intended results.

5

Sanitation
and the Small Farm
(1977)

In the time when my memories begin—the late 1930s—
people in the country did not go around empty-handed as
much as they do now. As I remember them from that time,
farm people on the way somewhere characteristically had
buckets or kettles or baskets in their hands, sometimes
sacks on their shoulders.

Those were hard times—not unusual in our agricultural
history—and so a lot of the fetching and carrying had to do
with foraging, searching the fields and woods for nature's
free provisions: greens in the springtime, fruits and berries
in the summer, nuts in the fall. There was fishing in warm
weather and hunting in cold weather; people did these
things for food and for pleasure, not for "sport." The
economies of many households were small and thorough,
and people took these seasonal opportunities seriously.

For the same reason, they practiced household husbandry. They raised gardens, fattened meat hogs, milked cows,
kept flocks of chickens and other poultry. These enterprises
were marginal to the farm, but central to the household. In
a sense, they comprised the direct bond between farm and
household. These enterprises produced surpluses which, in
those days, were marketable. And so when one saw farm

people in town they would be laden with buckets of cream or baskets of eggs. Or maybe you would see a woman going into the grocery store, carrying two or three old hens with their legs tied together. Sometimes this surplus paid for what the family had to buy at the store. Sometimes after they "bought" their groceries in this way, they had money to take home. These households were places of production, at least some of the time operating at a net economic *gain*. The idea of "consumption" was alien to them. I am not talking about practices of exceptional families, but about what was ordinarily done on virtually all farms.

That economy was in the truest sense democratic. Everybody could participate in it—even little children. An important source of instruction and pleasure to a child growing up on a farm was participation in the family economy. Children learned about the adult world by participating in it in a small way, by doing a little work and making a little money—a much more effective, because pleasurable, and a much cheaper method than the present one of requiring the adult world to be learned in the abstract in school. One's elders in those days were always admonishing one to save nickels and dimes, and there was tangible purpose in their advice: with enough nickels and dimes, one could buy a cow or a sow; with the income from a cow or a sow, one could begin to save to buy a farm. This scheme was plausible enough, evidently, for it seemed that all grown-ups had meditated on it. Now, according to the savants of agriculture—and most grown-ups now believe them—one does not start in farming with a sow or a cow; one must start with a quarter of a million dollars. What are the political implications of *that* economy?

I have so far mentioned only the most common small items of trade, but it was also possible to sell prepared foods: pies, bread, butter, beaten biscuits, cured hams, etc. And among the most attractive enterprises of that time were the small dairies that were added without much expense or trouble to the small, diversified farms. There would usually be a milking room or stall partitioned off in a barn, with homemade wooden stanchions to accommodate perhaps three to half a dozen cows. The cows were milked by hand. The milk was cooled in cans in a tub of well water. For a minimal expenditure and an hour or so of effort night and morning, the farm gained a steady, dependable income. All this conformed to the ideal of my grandfather's generation of farmers, which was to "sell something every week"—a maxim of diversity, stability, and small scale.

Both the foraging in fields and woods and the small husbandries of household and barn have now been almost entirely replaced by the "consumer economy," which assumes that it is better to buy whatever one needs than to find it or make it or grow it. Advertisements and other forms of propaganda suggest that people should congratulate themselves on the quantity and variety of their purchases. Shopping, in spite of traffic and crowds, is held to be "easy" and "convenient." Spending money gives one status. And physical exertion for any useful purpose is looked down upon; it is permissible to work hard for "sport" or "recreation," but to make any practical use of the body is considered beneath dignity.

Aside from the fashions of leisure and affluence—so valuable to corporations, so destructive of values—the greatest destroyer of the small economies of the small farms has been the doctrine of sanitation. I have no argument against cleanliness and healthfulness; I am for them as much as anyone. I do, however, question the validity and the honesty of the sanitation laws that have come to rule over farm production in the last thirty or forty years. Why have new sanitation laws always required more, and more expensive, equipment? Why have they always worked against the survival of the small producer? Is it impossible to be inexpensively healthful and clean?

I am not a scientist or a sanitation expert, and cannot give conclusive answers to those questions; I can only say what I have observed and what I think. In a remarkably short time I have seen the demise of all the small dairy operations in my part of the country, the shutting down of all local creameries and of all the small local dealers in milk and milk products. I have seen the grocers forced to quit dealing in eggs produced by local farmers, and have seen the closing of all markets for small quantities of poultry.

Recently, in continuation of this "trend," the local slaughterhouses in Kentucky were required to make expensive alterations or go out of business. Most of them went out of business. These were not offering meat for sale in the wholesale or retail trade. They did custom work mainly for local farmers who brought their animals in for slaughter and took the meat home or to a locker plant for processing. They were essential to the effort of many people to live self-sufficiently from their own produce—and these people had raised *no* objections to the way their meat was being handled. The few establishments that managed to survive this "improvement" found it necessary, of course, to charge higher prices for their work. Who benefitted from this? Not the customers, who were put to considerable expense and inconvenience, if they were not forced to quit producing their own meat altogether. Not, certainly, the slaughterhouses or the local economies. Not, so far as I can see, the public's health. The only conceivable beneficiaries were the meat-packing corporations, and for this questionable gain local life was weakened at its economic roots.

This sort of thing is always justified as "consumer protection." But we need to ask a few questions about that. How are consumers protected by a system that puts more and more miles, middlemen, agencies, and inspectors between them and the producers? How, over all these obstacles, can consumers make producers aware of their tastes and needs? How are consumers protected by a system that apparently cannot "improve" except by eliminating the small producer, increasing the cost of production, and increasing the retail price of the product?

Does the concentration of production in the hands of fewer and fewer big operators really serve the ends of cleanliness and health? Or does it make easier and more lucrative the possibility of collusion between irresponsible producers and corrupt inspectors?

In so strenuously and expensively protecting food from contamination by germs, how much have we increased the possibility of its contamination by antibiotics, preservatives, and various industrial poisons? The notorious PBB disaster in Michigan could probably not have happened in a decentralized system of small local suppliers and producers.

And, finally, what do we do to our people, our communities, our economy, and our political system when we allow our necessities to be produced by a centralized system of large operators, dependent on expensive technology, and regulated by expensive bureaucracy? The modern food industry is said to be a "miracle of technology." But it is well to remember that this technology, in addition to so-called miracles, produces economic and political consequences that are not favorable to democracy.

The connections among farming, technology, economics, and politics are important for many reasons, one of the most obvious being their influence on food production. Probably the worst fault of our present system is that it simply eliminates from production the land that is not suitable for, as well as the people who cannot afford, large-scale technology. And it ignores the potential productivity of these "marginal" acres and people.

It is possible to raise these issues because our leaders have been telling us for years that our agriculture needs to become more and more productive. If they mean what they say, they will have to revise production standards and open the necessary markets to provide a livelihood for small farmers. Only small farmers can keep the so-called marginal land in production, for only they can give the intensive care necessary to keep it productive.

6

Horse-Drawn Tools
and the Doctrine
of Labor Saving
(1978)

Five years ago, when we enlarged our farm from about
twelve acres to about fifty, we saw that we had come to the
limits of the equipment we had on hand: mainly a rotary
tiller and a Gravely walking tractor; we had been borrowing a tractor and mower to clip our few acres of pasture.
Now we would have perhaps twenty-five acres of pasture,
three acres of hay, and the garden; and we would also be
clearing some land and dragging the cut trees out for firewood. I thought for a while of buying a second-hand 8N
Ford tractor, but decided finally to buy a team of horses
instead.

I have several reasons for being glad that I did. One reason is that it started me thinking more particularly and
carefully than before about the development of agricultural
technology. I had learned to use a team when I was a boy,
and then had learned to use the tractor equipment that replaced virtually all the horse and mule teams in this part of
the country after World War II. Now I was turning around,
as if in the middle of my own history, and taking up the old
way again.

Buying and borrowing, I gathered up the equipment I
needed to get started: wagon, manure spreader, mowing

machine, disk, a one-row cultivating plow for the garden. Most of these machines had been sitting idle for years. I put them back into working shape, and started using them. That was 1973. In the years since, I have bought a number of other horse-drawn tools, for myself and other people. My own outfit now includes a breaking plow, a two-horse riding cultivator, and a grain drill.

As I have repaired these old machines and used them, I have seen how well designed and durable they are, and what good work they do. When the manufacturers modified them for use with tractors, they did not much improve either the machines or the quality of their work. (It is necessary, of course, to note some exceptions. Some horsemen, for instance, would argue that alfalfa sod is best plowed with a tractor. And one must also except such tools as hay conditioners and chisel plows that came after the development of horse-drawn tools had ceased. We do not know what innovations, refinements, and improvements would have come if it had continued.) At the peak of their development, the old horse tools were excellent. The coming of the tractor made it possible for a farmer to do more work, but not better. And there comes a point, as we know, when *more* begins to imply *worse*. The mechanization of farming passed that point long ago—probably, or so I will argue, when it passed from horse power to tractor power.

The increase of power has made it possible for one worker to crop an enormous acreage, but for this "efficiency" the country has paid a high price. From 1946 to 1976, because fewer people were needed, the farm population declined from thirty million to nine million; the rapid movement of these millions into the cities greatly aggravated that complex of problems which we now call the "urban crisis," and the land is suffering for want of the care of those absent families. The coming of a tool, then, can be a cultural event of great influence and power. Once that is understood, it is no longer possible to be simpleminded about technological progress. It is no longer possible to ask, What is a good tool? without asking at the same time, How *well* does it work? and, What is its influence?

One could say, as a rule of thumb, that a good tool is one that makes it possible to work faster *and* better than before. When companies quit making them, the horse-drawn tools fulfilled both requirements. Consider, for example, the International High Gear No. 9 mowing machine. This is a horse-drawn mower that certainly improved on everything that came before it, from the scythe to previous machines in the International line. Up to that point, to cut fast and to cut well were two aspects of the same problem. Past that point the speed of the work could be increased, but not the quality.

I own one of these mowers. I have used it in my hayfield at the same time that a neighbor mowed there with a tractor mower; I have gone from my own freshly cut hayfield into others just mowed by tractors; and I can say unhesitatingly that, though the tractors do faster work, they do not do it better. The same is substantially true, I think, of other tools: plows, cultivators, harrows, grain drills, seeders, spreaders, etc. Through the development of the standard horse-drawn equipment, quality and speed increased together; after that, the principle increase has been in speed.

Moreover, as the speed has increased, care has tended to decline. For this, one's eyes can furnish ample evidence. But we have it also by the testimony of the equipment manufacturers themselves. Here, for example, is a quote from the public relations paper of one of the largest companies: "Today we have multi-row planters that slap in a crop in a hurry, putting down seed, fertilizer, insecticide and herbicide in one quick swipe across the field."

But good work and good workmanship cannot be accomplished by "slaps" and "swipes." Such language seems to be derived from the he-man vocabulary of TV westerns, not from any known principles of good agriculture. What does the language of good agricultural workmanship sound like? Here is the voice of an old-time English farmworker and horseman, Harry Groom, as quoted in George Ewart Evans's *The Horse in the Furrow:* "It's all rush today. You hear a young chap say in the pub: 'I done thirty acres today.' But it ain't messed over, let alone done. You take the rolling, for instance. Two mile an hour is fast enough for a roll or a harrow. With a roll, the slower the better. If you roll fast, the clods are not broken up, they're just pressed in further. Speed is everything now; just jump on the tractor and way across the field as if it's a dirt-track. You see it when a farmer takes over a new farm: he goes in and plants straight-way, right out of the book. But if one of the old farmers took a new farm, and you walked round the land with him and asked him: 'What are you going to plant here and here?' he'd look at you some queer; because he wouldn't plant nothing much at first. He'd wait a bit and see what the land was like: he'd *prove* the land first. A good practical man would hold on for a few weeks, and get the feel of the land under his feet. He'd walk on it and feel it through his boots and see if it was in good heart, before he planted anything: he'd sow only when he knew what the land was fit for."

Granted that there is always plenty of room to disagree about farming methods, there is still no way to deny that in the first quotation we have a description of careless farming, and in the second a description of a way of farming as careful—as knowing, skillful, and loving—as any other kind of high workmanship. The difference between the two is simply that the second considers where and how the machine is used, whereas the first considers only the machine. The first is the point of view of a man high up in the air-conditioned cab of a tractor described as "a beast that eats acres." The second is that of a man who has worked close to the ground in the open air of the field, who has studied the condition of the ground as he drove over it, and who has cared and thought about it.

If we had tools thirty-five years ago that made it possible to do farm work both faster and better than before, then why did we choose to go ahead and make them no longer better, but just bigger and bigger and faster and faster? It was, I think, because we were already allowing the wrong people to give the wrong answers to questions raised by the improved horse-drawn machines. Those machines, like the ones that followed them, were *labor savers*. They may seem old-timey in comparison to today's "acre eaters," but when they came on the market they greatly increased the amount of work that one worker could do in a day. And so they confronted us with a critical question: How would we define labor saving?

We defined it, or allowed it to be defined for us by the corporations and the specialists, as if it involved no human considerations at all, as if the labor to be "saved" were not human labor. We decided, in the language of some experts, to look on technology as a "substitute for labor." Which means that we did not intend to "save" labor at all, but to *replace* it, and to *displace* the people who once supplied it. We never asked what should be done with the "saved" labor; we let the "labor market" take care of that. Nor did we ask the larger questions of what values we should place on people and their work and on the land. It appears that we abandoned ourselves unquestioningly to a course of technological evolution, which would value the development of machines far above the development of people.

And so it becomes clear that, by itself, my rule-of-thumb definition of a good tool (one that permits a worker to work both better and faster) does not go far enough. Even such a tool can cause bad results if its use is not directed by a benign and healthy social purpose. The coming of a tool, then, is not just a cultural event; it is also an historical crossroad—a point at which people must choose between two possibilities: to become more intensive or more extensive; to use the tool for quality or for quantity, for care or for speed.

In speaking of this as a choice, I am obviously assuming that the evolution of technology is *not* unquestionable or uncontrollable; that "progress" and the "labor market" do *not* represent anything so unyielding as natural law, but are aspects of an economy; and that any economy is in some sense a "managed" economy, managed by an intention to distribute the benefits of work, land, and materials in a certain way. (The present agricultural economy, for instance, is slanted to give the greater portion of these benefits to the "agribusiness" corporations. If this were not so, the recent farmers' strike would have been an "agribusiness" strike as well.) If those assumptions are correct, we are at liberty to do a little historical supposing, not meant, of course, to "change history" or "rewrite it," but to clarify somewhat this question of technological choice.

Suppose, then, that in 1945 we had valued the human life of farms and farm communities 1 percent more than we valued "economic growth" and technological progress. And suppose we had espoused the health of homes, farms, towns, and cities with anything like the resolve and energy with which we built the "military-industrial complex." Suppose, in other words, that we had really meant what, all that time, most of us and most of our leaders were saying, and that we had really tried to live by the traditional values to which we gave lip service.

Then, it seems to me, we might have accepted certain mechanical and economic limits. We might have used the improved horse-drawn tools, or even the small tractor equipment that followed, not to displace workers and decrease care and skill, but to intensify production, improve maintenance, increase care and skill, and widen the margins of leisure, pleasure, and community life. We might, in other words, by limiting technology to a human or a democratic scale, have been able to use the saved labor *in the same places where we saved it.*

It is important to remember that "labor" is a very crude, industrial term, fitted to the huge economic structures, the dehumanized technology, and the abstract social organization of urban-industrial society. In such circumstances, "labor" means little more than the sum of two human quantities, human energy plus human time, which we identify as "man-hours." But the nearer home we put "labor" to work, and the smaller and more familiar we make its circumstances, the more we enlarge and complicate and enhance its meaning. At work in a factory, workers are only workers, "units of production" expending "man-hours" at a task set for them by strangers. At work in their own communities, on their own farms or in their own households or shops, workers are *never* only workers, but rather persons, relatives, and neighbors. They work *for* those they work *among* and *with.* Moreover, workers tend to be independent in inverse proportion to the size of the circumstance in which they work. That is, the work of factory workers is ruled by the factory, whereas the work of housewives, small craftsmen, or small farmers is ruled by their own morality, skill, and intelligence. And so, when workers work independently and at home, the society as a whole may lose something in the way of organizational efficiency and economies of scale. But it begins to *gain* values not so readily quantifiable in the fulfilled humanity of the workers, who then bring to their work not just contracted quantities of "man-hours," but qualities such as independence, skill, intelligence, judgment, pride, respect, loyalty, love, reverence.

To put the matter in concrete terms, if the farm communities had been able to use the best horse-drawn tools to save labor in the true sense, then they might have used the saved time and energy, first of all, for leisure—something that technological progress has given to farmers. Second, they might have used it to improve their farms: to enrich the soil, prevent erosion, conserve water, put up better and more permanent fences and buildings; to practice forestry and its dependent crafts and economies; to plant orchards, vineyards, gardens of bush fruits; to plant market gardens; to improve pasture, breeding, husbandry, and the subsidiary enterprises of a local, small-herd livestock economy; to enlarge, diversify, and deepen the economies of households and homesteads. Third, they might have used it to expand and improve the specialized crafts necessary to the health and beauty of communities: carpentry, masonry, leatherwork, cabinetwork, metalwork, pottery, etc. Fourth, they might have used it to improve the homelife and the home instruction of children, thereby preventing the hardships and expenses now placed on schools, courts, and jails.

It is probable also that, if we *had* followed such a course, we would have averted or greatly ameliorated the present shortages of energy and employment. The cities would be much less crowded; the rates of crime and welfare dependency would be much lower; the standards of industrial production would probably be higher. And farmers might have avoided their present crippling dependence on money lenders.

I am aware that all this is exactly the sort of thinking that the technological determinists will dismiss as nostalgic or wishful. I mean it, however, not as a recommendation that we "return to the past," but as a criticism of the past; and my criticism is based on the assumption that we had in the past, and that we have now, a *choice* about how we should use technology and what we should use it for. As I understand it, this choice depends absolutely on our willingness to limit our desires as well as the scale and kind of technology we use to satisfy them. Without that willingness, there is no choice; we must simply abandon ourselves to whatever the technologists may discover to be possible.

The technological determinists, of course, do not accept that such a choice exists—undoubtedly because they resent the moral limits on their work that such a choice implies. They speak romantically of "man's destiny" to go on to bigger and more sophisticated machines. Or they take the opposite course and speak the tooth-and-claw language of Darwinism. Ex-secretary of agriculture Earl Butz speaks, for instance, of "Butz's Law of Economics" which is "Adapt or Die."

I am, I think, as enthusiastic about the principle of adaptation as Mr. Butz. We differ only on the question of what should be adapted. He believes that we should adapt to the machines, that humans should be forced to conform to technological conditions or standards. I believe that the machines should be adapted to us—to serve our *human* needs as our history, our heritage, and our most generous hopes have defined them.

7

Agricultural Solutions
for Agricultural Problems
(1978)

It may turn out that the most powerful and the most destructive change of modern times has been a change in language: the rise of the image, or metaphor, of the machine. Until the industrial revolution occurred in the minds of most of the people in the so-called "developed" countries, the dominant images were organic: they had to do with living things; they were biological, pastoral, agricultural, or familial. God was seen as a "shepherd," the faithful as "the sheep of His pasture." One's home country was known as one's "motherland." Certain people were said to have the strength of a lion, the grace of a deer, the speed of a falcon, the cunning of a fox, etc. Jesus spoke of himself as a "bridegroom." People who took good care of the earth were said to practice "husbandry." The ideal relationships among people were "brotherhood" and "sisterhood."

Now we do not flinch to hear men and women referred to as "units" as if they were as uniform and interchangeable as machine parts. It is common, and considered acceptable, to refer to the mind as a computer: one's thoughts are "inputs:" other people's responses are "feedback." And the body is thought of as a machine; it is said, for instance, to use food as "fuel;" and the best workers and athletes are praised by being compared to machines. Work is judged almost exclusively now by its "efficiency," which, as used, is a mechanical standard, or by its profitability, which is our only trusted index of mechanical efficiency. One's country is no longer loved familially and intimately as a "motherland," but rather priced according to its "productivity" of "raw materials" and "natural resources"—valued, that is, strictly according to its ability to keep the machines running. And recently R. Buckminster Fuller asserted that "the universe physically is itself the most incredible technology"—the necessary implication being that God is not father, shepherd, or bridegroom, but a mechanic, operating by principles which, according to Fuller, "can only be expressed mathematically."

In view of this revolution of language, which is in effect the uprooting of the human mind, it is not surprising to realize that farming too has been made to serve under the yoke of this extremely reductive metaphor. Farming, according to most of the most powerful people now concerned with it, is no longer a way of life, no longer husbandry or even agriculture; it is an industry known as "agribusiness," which looks upon a farm as a "factory," and upon farmers, plants, animals, and the land itself as interchangeable parts or "units of production."

This view of farming has been dominant now for a generation, and so it is not too soon to ask: How well does it work? We must answer that it works as any industrial machine works: very "efficiently" according to the terms of an extremely specialized accounting. That is to say that it *apparently* makes it possible for about 4 percent of the population to "feed" the rest. So long as we keep the focus narrowed to the "food factory" itself, we have to be impressed: it is elaborately organized; it is technologically sophisticated; it is, by its own definition of the term, marvelously "efficient."

Only when we widen the focus do we see that this "factory" is in fact a failure. Within itself it has the order of a machine, but, like other enterprises of the industrial vision, it is part of a rapidly widening and deepening disorder. It will be sufficient here to list some of the serious problems that have a demonstrable connection with industrial agriculture: (1) soil erosion, (2) soil compaction, (3) soil and water pollution, (4) pests and diseases resulting from monoculture and ecological deterioration, (5) depopulation of rural communities, and (6) decivilization of the cities.

The most obvious falsehood of "agribusiness" accounting has to do with the alleged "efficiency" of "agribusiness" technology. This is, in the first place, an efficiency calculated in the productivity of workers, not of acres. In the second place the productivity per "man-hour," as given out by "agribusiness" apologists, is dangerously—and, one must assume, intentionally—misleading. For the 4 percent of our population that is left on the farm does not, by any stretch of imagination, feed the rest. That 4 percent is only a small part, and the worst-paid part, of a food production network that includes purchasers, wholesalers, retailers, processors, packagers, transporters, and the manufacturers and salesmen of machines, building materials, feeds, pesticides, herbicides, fertilizers, medicines, and fuel. All these producers are at once in competition with each other and dependent on each other, and all are dependent on the petroleum industry.

As for the farmers themselves, they have long ago lost control of their destiny. They are no longer "independent farmers," subscribing to that ancient and perhaps indispensable ideal, but are agents of their creditors and of the market. They are "units of production" who, or *which*, must perform "efficiently"—regardless of what they get out of it either as investors or as human beings.

In the larger accounting, then, industrial agriculture is a failure on its way to being a catastrophe. Why is it a failure? There are, I think, two inescapable reasons.

The first is that the industrial vision is perhaps inherently an oversimplifying vision, which proceeds on the assumption that consequence is always singular; industrialists invariably assume that they are solving for X—X being production. In order to solve for X, industrial agriculturists have to reduce any agricultural problem to a problem in mechanics—as, for example, modern confinement-feeding techniques became possible only when animals could be considered as machines.

What this vision excludes, as a matter of course, are biology on the one hand, and human culture on the other. Once vision is enlarged to include these considerations, we see readily that—as wisdom has always counseled us—consequences are invariably multiple, self-multiplying, long lasting, and unforeseeable in something like geometric proportion to the size or power of the cause. Taking our bearings from traditional wisdom and from the insights of the ecologists—which, so far as I can see, confirm traditional wisdom—we realize that in a country the size of the United States, and economically uniform, the smallest possible agricultural "unit of production" is very large indeed. It consists of all the farmland, plus all the farmers, plus all the farming communities, plus all the knowledge and the technical means of agriculture, plus all the available species of domestic plants and animals, plus the natural systems and cycles that surround farming and support it, plus the knowledge, taste, judgment, kitchen skills, etc. of all the people who buy food. A proper solution to an agricultural problem must preserve and promote the good health of this "unit." Nothing less will do.

The second reason for the failure of industrial agriculture is its wastefulness. In natural or biological systems, waste does not occur. And it is easy to produce examples of nonindustrial human cultures in which waste was or is virtually unknown. All that is sloughed off in the living arc of a natural cycle remains within the cycle; it becomes fertility, the power of life to continue. In nature death and decay are as necessary—are, one may almost say, as lively—as life; and so nothing is wasted. There is really no such thing, then, as natural production; in nature, there is only reproduction.

But waste—so far, at least—has always been intrinsic to industrial production. There have always been unusable "by-products." Because industrial cycles are never complete—because there is no *return*—there are two characteristic results of industrial enterprise: exhaustion and contamination. The energy industry, for instance, is not a cycle, but only a short arc between an empty hole and poisoned air. And farming, which is inherently cyclic, capable of regenerating and reproducing itself indefinitely, becomes similarly destructive and self-exhausting when transformed into an industry. Agricultural pollution is a serious and growing problem. And industrial agriculture is forced by its very character to treat the soil itself as a "raw material," which it proceeds to "use up." It has been estimated, for instance, that at the present rate of cropland erosion Iowa's soil will be exhausted by the year 2050. I have seen no attempt to calculate the *human* cost of such farming—by attrition, displacement, social disruption, etc.—I assume because it is incalculable.

This failure of industrial agriculture is not more obvious, or more noticed, because many of its worst social and economic consequences have collected in the cities, and are erroneously called "urban problems." Also, because the farm population is now so small, most people know nothing of farming, and cannot recognize agricultural problems when they see them.

But if industrial agriculture is a failure, then how does it continue to produce such an enormous volume of food? One reason is that most countries where industrial agriculture is practiced have soils that were originally good, possessing great natural reserves of fertility. (Industrial agriculture is much more quickly destructive in places where the fertility reserves of the soil are not great—as in the Amazon basin.) Another reason is that, as natural fertility has declined, we have so far been able to subsidize food production by large applications of chemical fertilizer. These have effectively disguised the loss of natural fertility, but it is important to emphasize that they *are* a disguise. They delay some of the consequences of failure, but cannot prevent them. Chemical fertilizers are required in vast amounts, they are increasingly expensive, and most of them come from sources that are not renewable. Industrial agriculture is now absolutely dependent on them, and this dependence is one of its fundamental weaknesses.

Another weakness of industrial agriculture is its absolute dependence on an enormous and intricate—hence fragile—economic and industrial organization. Industrial food production can be gravely impaired or stopped by any number of causes, none of which need be agricultural: a truckers' strike, an oil shortage, a credit shortage, a manufacturing "error" such as the PBB catastrophe in Michigan.

A third weakness is the absolute dependence of most of the population on industrial agriculture—and the lack of any "backup system." We have an unprecedentedly large urban population that has no land to grow food on, no knowledge of how to grow it, and less and less knowledge of what to do with it after it is grown. That this population can continue to eat through shortage, strike, embargo, riot, depression, war—or any of the other large-scale afflictions that societies have always been heir to and that industrial societies are uniquely vulnerable to—is not a certainty or even a faith; it is a superstition.

As an example of the unexamined confusions and contra-dictions that underlie industrial agriculture, consider Agri-culture Secretary Bob Bergland's recent remarks on the state of agriculture in China: "From the manpower-production point of view, they're terribly inefficient—700 million people doing the most pedestrian kind of things. But in production per acre, they're enormously successful. They get nine times as many calories per acre as we do in the United States."

This comment is remarkable for its failure to acknowl-edge any possible connection between China's large agri-cultural work force and its high per-acre productivity. In many parts of China, according to one recent observer, the agriculture is still much closer to what we call gardening than to what we call farming. Because their farming is done on comparatively small plots, using a lot of hand labor, Chinese farmers have at their disposal such high-pro-duction techniques as intercropping and close rotations, which with us are available only to home gardeners. Many Chinese fields have maintained the productivity of gardens for thousands of years, and this is directly attributable to the great numbers of the farming population. Each acre can be intensively used and cared for, maintained for cen-turies at maximum fertility and yield, because there are enough knowledgeable people to do the necessary hand-work.

It is naive to assume, as Mr. Bergland implicitly does, that such an agriculture can be improved by "moderniza-tion"—that is, by the introduction of industrial standards, methods, and technology. How can this agriculture be in-dustrialized without destroying its intensive methods, and thus reducing its productivity per acre? How can the so-called "pedestrian" tasks be taken over by machines with-out displacing people, increasing unemployment, degrad-ing the quality of land maintenance, increasing slums and other urban blights? How, in other words, can this revolution fail to cause in China the same disorders that it has already caused in the United States? I do not mean to imply that these questions can be answered simply. My point is that before we participate in the industrialization of Chinese agriculture we ought to ask and answer these questions.

Finally, the Secretary's statement is remarkable for its revealing use of the word "pedestrian." This is a usage strictly in keeping with the industrial revolution of our language. The farther industrialization has gone with us, and the more it has influenced our values and behavior, the more contemptuous and belittling has the adjective "pedestrian" become. If you want to know how highly anything "pedestrian" is regarded, try walking along the edge of a busy highway; you will see that you are regarded mainly as an obstruction to the progress of greater power and velocity. The less power and velocity a thing has, the more "pedestrian" it is. A plow with one bottom is, as a matter of course, more "pedestrian" than a plow with eight bottoms; the quality of use is not recognized as an issue. The hand laborers are thus to be eliminated from China's fields for the same reason that we now build housing developments without sidewalks: the pedestrian, not being allowed *for*, is not allowed. By the use of this term, the Secretary ignores the issue of the quality of work on the one hand, and on the other hand the issue of social values and aims. Is field work necessarily improved when done with machines instead of people? And is a worker necessarily improved by being replaced by a machine? Does a worker invariably work better, more ably, with more interest and satisfaction when his power is mechanically magnified? And is a worker better off working at a "pedestrian" farm task or unemployed in an urban ghetto? In which instance is his country better off?

I have belabored Secretary Bergland's statement at such

length not because it is so odd, but because it is so characteristic of the dominant American approach to agriculture. He is using—unconsciously, I suspect—the language of agricultural industrialism, which fails to solve agricultural problems correctly because it cannot understand or define them as agricultural problems.

I will try now to define an approach to agriculture that is agricultural, that will lead to proper solutions, and that will, in consequence, safeguard and promote the health of the great unit of food production, which includes us all and all of our country. In order to do this I will deal with four problems, which seem to me inherent in the discipline of farming, and which are practical in the sense that their ultimate solutions cannot occur in public places—in organizations, in markets, or in policies—but only *on farms*. These are the problems of scale, of balance, of diversity, of quality. That these problems cannot be separated, and that no one of them can be solved without solving the others, testifies to their authenticity.

    1. The Problem of Scale. The identification of scale as a "problem" implies that things can be too big as well as too small, and I believe that this is so. Technology can grow to a size that is first undemocratic and then inhuman. It can grow beyond the control of individual human beings—and so, perhaps, beyond the control of human institutions. How large can a machine be before it ceases to serve people and begins to subjugate them?

    The size of land holdings is likewise a *political* fact. In any given region there is a farm size that is democratic, and a farm size that is plutocratic or totalitarian. A great danger to democracy now in the United States is the steep decline in the number of people who own farmland—or landed property of any kind. (According to a just-published report of the General Accounting Office, "To-day, it is estimated that less than one-half of all farmland is owned by the operator.") Earl Butz has suggested that this is made up for by the increased numbers of people who own insurance policies. But the value of insurance policies fluctuates with the value of money, whereas the *real* value of land never varies; it is always equal to the value of survival, of life. When this value is controlled by a wealthy or powerful minority, then democracy is reduced to mere governmental *forms,* easy to destroy or ignore.

Moreover, in any given region there is a limit beyond which a farm outgrows the attention, affection, and care of a single owner.

The size of fields is also a matter of agricultural concern. Fields can be too big to permit effective rotation of grazing, or to prevent erosion of land in cultivation. In general, the steeper the ground, the smaller should be the fields. On the steep slopes of the Andes, for instance, agriculture has survived for thousands of years. This survival has obviously depended on holding the soil in place, and the Andean peasants have an extensive methodology of erosion control. Of all their means and methods, none is more important than the smallness of their fields—which is permitted by the smallness of their technology, most of the land still being worked by hand or with oxen.

2. The Problem of Balance. Finding the correct ratio between people and land, so that maintenance always equals production. This is obviously related to the problem of scale. In the correct solution to these problems, such problems as soil erosion and soil compaction will be solved.

But also each farm and each farmer must establish the proper ratio between plants and animals. This is the foundation of agricultural independence. In this balance of plants and animals the fertility cycle is kept complete, or as nearly complete as possible. Ideally, the farm would provide its own fertility. However, in commercial farming, when so many nutrients are shipped off the farm as food, it is necessary to return them to the farm in the form of composted "urban wastes"—sewage, garbage, etc.

By studying the problem of balance, one discovers the carrying capacity of a farm—that is, the amount it can produce without diminishing its ability to produce.

When the problem of balance is solved, a farm's production becomes more or less constant. The farm will no longer be stocked or cultivated according to fluctuations of the market—which is not agriculture but an imitation, on the farm, of industrial economics.

3. The Problem of Diversity. This is the only possible agricultural "backup system." On the farm it means not putting all the eggs in one basket; it means—within the limits of nature, sense, and practicality—having as many kinds, as many species, as possible.

In terms of our country's agriculture as a whole, too, it means the diversity of species. But it also means as many different kinds of *good* agriculture as possible: farms changing in kind, as necessary, from one location to another; but also truck farms and part-time farms near cities, to increase local self-sufficiency and independence; and home gardens everywhere, in the cities as well as in the country.

4. The Problem of Quality. *Quality,* as I shall understand it here, is indistinguishable from *health*—bodily health, coming from good food, but also economic, political, cultural, and spiritual health. All these kinds of health are related. And I hope that my discussion of the other problems has begun to make clear how dependent health is on good work.

Industrial agriculture has tended to look on the farmer as a "worker"—a sort of obsolete but not yet dispensable machine—acting on the advice of scientists and economists. We have neglected the truth that a *good* farmer is a craftsman of the highest order, a kind of artist. It is the good work of good farmers—nothing else—that assures a sufficiency of food over the long term.

Ignoring that, industrial economics has encouraged poor work on the farm. I believe that it has done so because poor work can be easily priced. Since poor work lasts only a short time, the money value of its whole life can be readily calculated. Good work, which in fact or influence endures beyond the foresight of economists, can be valued but not priced, because its worth is incalculable. I am talking about the difference, say, between a wire fence and a stone wall, or between any gasoline engine and any good breed of live-stock.

I am more and more convinced that the only guarantee of quality in practice lies in the subsistence principle—that is, in the use of the product by the producer—a principle depreciated virtually out of existence by industrial agriculture. Indeed, it is sometimes offered as one of the benefits of industrial agriculture that farm families now patronize the supermarkets just like city people. On the other hand, it can be well argued that people who use their own products will be as concerned for quality as for quantity, whereas people who produce exclusively for the market will be mainly interested in quantity.

It will be noticed that production is not on my list of problems. The reason is that if the four problems I have dealt with are properly solved, production will not be a problem. Good production is merely the result of good farming.

8

Energy in Agriculture
(1979)

I have just been rereading Donald Hall's lovely memoir, *String Too Short to Be Saved*. It is about the summers of his boyhood that the author spent on his grandparents' New Hampshire farm, from the late 1930s until the early 1950s. There are many good things in this book, but one of the best is its description of the life and economy of an old-time New England small farm.

The farm of Kate and Wesley Wells, as their grandson knew it, was already a relic. It was what would now be called a "marginal farm" in mountainous country, in an agricultural community that had been dying since the Civil War. The farm produced food for the household and made a cash income from a small hand-milked herd of Holsteins and a flock of sheep. It furnished trees for firewood and maple syrup. The Wellses sent their daughters to school by the sale of timber from a woodlot. The farm and its household were "poor" by our present standards, taking in very little money—but spending very little too, and that is the most important thing about it. Its principle was thrift. Its needs were kept within the limits of its resources.

This farm was ordered according to an old agrarian pattern which made it far more independent than modern farms built upon the pattern of industrial capitalism. And its energy economy was as independent as its money economy. The working energy of this farm came mainly from its people and from one horse.

Mr. Hall's memories inform us, more powerfully than any argument, that the life of Wesley and Kate Wells was a life worth living, decent though not easy; not adventurous or affluent, either—or not in our sense—but sociable, neighborly, and humane. They were intelligent, morally competent, upright, kind to people and animals, full of generous memories and good humor. From all that their grandson says of them, it is clear that his acquaintance with them and their place was profoundly enabling to his mind and his feelings.

One cannot read this book—or I, anyhow, cannot—without asking how that sort of life escaped us, how it depreciated as a possibility so that we were able to give it up in order, as we thought, to "improve" ourselves. Mr. Hall makes it plain that farms like his grandparents' did not die out in New England necessarily because of bad farming, or because they did not provide a viable way of life. They died for want of people with the motivation, the skill, the character, and the culture to keep them alive. They died, in other words, by a change in cultural value. Though it survived fairly intact until the middle of this century, Mr. Hall remembers that his grandparents' farm was surrounded by people and farms that had dwindled away because the human succession had been broken. It was no longer a place to come to, but a place to leave.

At the time Mr. Hall writes about, something was gaining speed in our country that I think will seem more and more strange as time goes on. This was a curious set of assumptions, both personal and public, about "progress." If you could get into a profession, it was assumed, then of course you must not be a farmer; if you could move to the city, then you must not stay in the country; if you could farm more profitably in the corn belt than on the mountainsides of New England, then the mountainsides of New England must not be farmed. For years this set of assumptions was rarely spoken and more rarely questioned, and yet it has been one of the most powerful social forces at work in this country in modern times.

But these assumptions could not accomplish much on their own. What gave them power, and made them able finally to dominate and reshape our society, was the growth of technology for the production and use of fossil fuel energy. This energy could be made available to empower such unprecedented social change because it was "cheap." But we were able to consider it "cheap" only by a kind of moral simplicity: the assumption that we had a "right" to as much of it as we could use. This was a "right" made solely by might. Because fossil fuels, however abundant they once were, were nevertheless limited in quantity and not renewable, they obviously did not "belong" to one generation more than another. We ignored the claims of posterity simply because we could, the living being stronger than the unborn, and so worked the "miracle" of industrial progress by the theft of energy from (among others) our children.

That is the real foundation of our progress and our affluence. The reason that we are a rich nation is not that we have earned so much wealth—you cannot, by any honest means, earn or deserve so much. The reason is simply that we have learned, and become willing, to market and use up in our own time the birthright and livelihood of posterity.

And so it is too simple to say that the "marginal" farms of New England were abandoned because of progress or because they were no longer productive or desirable as living places. They were given up for one very "practical" reason: they did not lend themselves readily to exploitation by fossil fuel technology. Their decline began with the rise of steam power and the industrial economy after the Civil War; the coming of industrial agriculture after World War II finished them off. Industrial agriculture needs large holdings and large level fields. As the scale of technology grows, the small farms with small or steep fields are pushed farther and farther toward the economic margins and are finally abandoned. And so industrial agriculture sticks itself deeper and deeper into a curious paradox: the larger its technology grows in order "to feed the world," the more potentially productive "marginal" land it either ruins or causes to be abandoned. If the sweeping landscapes of Nebraska now have to be reshaped by computer and bulldozer to allow the more efficient operation of big farm machines, then thousands of acres of the smaller-featured hill country of the eastern states must obviously be considered "unfarmable." Or so the industrialists of agriculture have ruled.

And so energy is not just fuel. It is a powerful social and cultural influence. The kind and quantity of the energy we use determine the kind and quality of the life we live. Our conversion to fossil fuel energy subjected society to a sort of technological determinism, shifting population and values according to the new patterns and values of industrialization. Rural wealth and materials and rural people were caught within the gravitational field of the industrial economy and flowed to the cities, from which comparatively little flowed back in return. And so the human life of farmsteads and rural communities dwindled everywhere, and in some places perished.

If the shift to fossil fuel energy radically changed the life and the values of farm communities, it should be no surprise that it also radically changed our understanding of

agriculture. Some figures from an article by Professor
Mark D. Shaw help to show the nature of this change. The
"food system," according to Professor Shaw, now uses 16.5
percent of all energy used in the United States. This 16.5
percent is used in the following ways:

On-farm production	3.0%
Manufacturing	4.9
Wholesale marketing	0.5
Retail marketing	0.8
Food preparation (in home)	4.4
Food preparation (commercial)	2.9

Apologists for industrial agriculture frequently stop with
that first figure—showing that agriculture uses only a small
amount of energy, relatively speaking, and that people
hunting a cause of the "energy crisis" should therefore
point their fingers elsewhere. The other figures, amounting
to 13.5 percent of national energy consumption, are more
interesting, for they suggest the way the food system has
been expanded to make room for industrial enterprise. Between farm and home, producer and consumer, we have in-
terposed manufacturers, a complex marketing structure,
and food preparation. I am not sure how this last category
differs from "manufacturing." And I would like to know
what percentage of the energy budget goes for transportation, and whether or not Professor Shaw figured in the
miles that people now drive to shop. The gist is nevertheless
plain enough: the industrial economy grows and thrives by
lengthening and complicating the essential connection between producer and consumer. In a *local* food economy,
dealing in fresh produce to be prepared in the home (thus
eliminating transporters, manufacturers, packagers, pre-
parers, etc.), the energy budget would be substantially lower, and we might have both cheaper food and higher earnings on the farm.

But Professor Shaw provides another set of figures that is

even more telling. These have to do with the "sources of energy for Pennsylvania agriculture" (I don't think the significance would vary much from one state to another):

Nuclear 1%
Coal 5
Natural gas 27
Petroleum 67

And so we see that, though our agriculture may use relatively little fossil fuel energy, it is almost totally dependent on what it does use. It uses fossil fuel energy almost exclusively and uses it in competition with other users. And the sources of this energy are not renewable.

This critical dependence on nonrenewable energy sources is the direct result of the industrialization of agriculture. Before industrialization, agriculture depended almost exclusively on solar energy. Solar energy not only grew the plants, as it still does, but also provided the productive power of farms in the form of the work of humans and animals. This energy is derived and made available biologically, and it is recyclable. It is inexhaustible in the topsoil so long as good husbandry keeps the life cycle intact.

This old sun-based agriculture was fundamentally alien to the industrial economy; industrial corporations could make relatively little profit from it. In order to make agriculture fully exploitable by industry it was necessary (in Barry Commoner's terms) to weaken "the farm's link to the sun" and to make the farmland a "colony" of the industrial corporations. The farmers had to be persuaded to give up the free energy of the sun in order to pay dearly for the machine-derived energy of the fossil fuels.

Thus we have another example of a system artificially expanded for profit. The farm's originally organic, coherent, independent production system was expanded into a complex dependence on remote sources and on manufactured supplies.

What happened, from a cultural point of view, was that

machines were substituted for farmers, and energy took the place of skill. As farmers became more and more dependent on fossil fuel energy, a radical change occurred in their minds. Once focused on biology, the life and health of living things, their thinking now began to focus on technology and economics. Credit, for example, became as pressing an issue as the weather, for farmers had begun to climb the one-way ladder of survival by debt. Bigger machines required more land, and more land required yet bigger machines, which required yet more land, and on and on—the survivors climbing to precarious and often temporary success by way of machines and mortgages and the ruin of their neighbors. And so the farm became a "factory," where speed, "efficiency," and profitability were the main standards of performance. These standards, of course, are industrial, not agricultural.

The old solar agriculture, moreover, was time oriented. Timeliness was its virtue. One took pride in having the knowledge to do things at the right time. Industrial agriculture is space oriented. Its virtue is speed. One takes pride in being first. The right time, by contrast, could be late as well as early; the proof of the work was in its quality.

The most important point I have to make is that once agriculture shifted its dependence from solar, biologically derived energy to machine-derived fossil fuel energy, it committed itself, as a matter of course, to several kinds of waste:

1. The waste of solar energy, not just as motive power, but even as growing power. As landholdings become larger and the number of farmers smaller, more and more fields must go without cover crops, which means that for many days in the fall and early spring the sunlight on these fields is not captured in green leaves and so made useful to the soil and to people. It goes to waste.

2. The waste of human energy and ability. Industrial agriculture replaces people with machines; the ability of millions of people to become skillful and to do work therefore comes to nothing. We now have millions on some kind of government support, grown useless and helpless, while our country becomes unhealthy and ugly for want of human work and care. And we have additional millions not on welfare who have grown almost equally useless and helpless for want of health. How much potentially useful energy do we now have stored in human belly fat? And what is it costing us, not only in medical bills, but in money spent on diets, drugs, and exercise machines?

3. The waste of animal energy. I mean not just the abandonment of live horsepower, but the waste involved in confinement feeding. Why use fossil fuel energy to bring food to grazing animals that are admirably designed to go get it themselves?

4. The waste of soil and soil health. Because the number of farmers has now grown so small in proportion to the number of acres that must be farmed, it has been necessary to resort to all sorts of mechanical shortcuts. But shortcuts never have resulted in good work, and there is no reason to believe that they ever will. When a farmer must cover an enormous acreage within the strict limits of the seasons of planting and harvest, speed necessarily becomes the first consideration. And so the machinery, not the land, becomes the focus of attention and the standard of the work. Consequently, the fields get larger so as to require less turning, waterways are plowed out, and one sees less and less terracing and contour or strip plowing. And, as I mentioned above, less and less land is sowed in a cover crop; when such large acreages must be harvested, there is no time for a fall seeding. The result is catastrophic soil erosion even in such "flat" states as Iowa.

A problem related to soil waste is that of soil compaction. Part of the reason for this is that industrial agriculture reduces the humus in the soil, which becomes more cohesive and less porous as a result. Another reason is the use of heavier equipment, which becomes necessary, in the first place, because of soil compaction. But the main reason, I think, is again that we don't have enough farmers to farm the land properly. The industrial farmer has so much land that he cannot afford to wait for "the right time" to work his fields. As long as the ground will support his equipment, he plows and harrows; the time is right for the work whenever the work is mechanically possible. It is commonplace now, wherever I have traveled in farm country, to see fields cut to pieces by deep wheel tracks.

The final irony is that we are abusing our land in this way partly in order to correct our "balance of payments" —that is, in order to buy foreign petroleum. In the language of some "agribusiness" experts we are using "agridollars" to offset the drain of "petrodollars." We are, in effect, exporting our topsoil in order to keep our tractors running.

There is no question that you can cover a lot of ground with the big machines now on the market. A lot of people seem entranced by the power and speed of those machines, which the manufacturers love to refer to as "monsters" and "acre eaters." But the result is not farming; it is a process closely akin to mining. In what is left of the country communities, in earshot of the monster acre eaters of the "agribusinessmen," a lot of old farmers must be turning over in their graves.

9

Solving for Pattern

(1980)

Our dilemma in agriculture now is that the industrial methods that have so spectacularly solved some of the problems of food production have been accompanied by "side effects" so damaging as to threaten the survival of farming. Perhaps the best clue to the nature and the gravity of this dilemma is that it is not limited to agriculture. My immediate concern here is with the irony of agricultural methods that destroy, first, the health of the soil and, finally, the health of human communities. But I could just as easily be talking about sanitation systems that pollute, school systems that graduate illiterate students, medical cures that cause disease, or nuclear armaments that explode in the midst of the people they are meant to protect. This is a kind of surprise that is characteristic of our time: the cure proves incurable; security results in the evacuation of a neighborhood or a town. It is only when it is understood that our agricultural dilemma is characteristic not of our agriculture but of our time that we can begin to understand why these surprises happen, and to work out standards of judgment that may prevent them.

To the problems of farming, then, as to other problems of our time, there appear to be three kinds of solutions:

There is, first, the solution that causes a ramifying series of new problems, the only limiting criterion being, apparently, that the new problems should arise beyond the purview of the expertise that produced the solution—as, in agriculture, industrial solutions to the problem of production have invariably caused problems of maintenance, conservation, economics, community health, etc., etc.

If, for example, beef cattle are fed in large feed lots, within the boundaries of the feeding operation itself a certain factory-like order and efficiency can be achieved. But even within those boundaries that mechanical order immediately produces biological disorder, for we know that health problems and dependence on drugs will be greater among cattle so confined than among cattle on pasture.

And beyond those boundaries, the problems multiply. Pen feeding of cattle in large numbers involves, first, a manure-removal problem, which becomes at some point a health problem for the animals themselves, for the local watershed, and for the adjoining ecosystems and human communities. If the manure is disposed of without returning it to the soil that produced the feed, a serious problem of soil fertility is involved. But we know too that large concentrations of animals in feed lots in one place tend to be associated with, and to promote, large cash-grain monocultures in other places. These monocultures tend to be accompanied by a whole set of specifically agricultural problems: soil erosion, soil compaction, epidemic infestations of pests, weeds, and disease. But they are also accompanied by a set of agricultural-economic problems (dependence on purchased technology; dependence on purchased fuels, fertilizers, and poisons; dependence on credit)—and by a set of community problems, beginning with depopulation and the removal of sources, services, and markets to more and more distant towns. And these are, so to speak, only the first circle of the bad effects of a bad solution. With a little care, their branchings can be traced on into nature, into the life of the cities, and into the cultural and economic life of the nation.

The second kind of solution is that which immediately worsens the problem it is intended to solve, causing a hellish symbiosis in which problem and solution reciprocally enlarge one another in a sequence that, so far as its own logic is concerned, is limitless—as when the problem of soil compaction is "solved" by a bigger tractor, which further compacts the soil, which makes a need for a still bigger tractor, and so on and on. There is an identical symbiosis between coal-fired power plants and air conditioners. It is characteristic of such solutions that no one prospers by them but the suppliers of fuel and equipment.

These two kinds of solutions are obviously bad. They always serve one good at the expense of another or of several others, and I believe that if all their effects were ever to be accounted for they would be seen to involve, too frequently if not invariably, a net loss to nature, agriculture, and the human commonwealth.

Such solutions always involve a definition of the problem that is either false or so narrow as to be virtually false. To define an agricultural problem as if it were solely a problem of agriculture—or solely a problem of production or technology or economics—is simply to misunderstand the problem, either inadvertently or deliberately, either for profit or because of a prevalent fashion of thought. The whole problem must be solved, not just some handily identifiable and simplifiable aspect of it.

Both kinds of bad solutions leave their problems unsolved. Bigger tractors do not solve the problem of soil compaction any more than air conditioners solve the problem of air pollution. Nor does the large confinement-feeding operation solve the problem of food production; it is, rather, a way calculated to allow large-scale ambition and greed to profit from food production. The real problem of food production occurs within a complex, mutually influential relationship of soil, plants, animals, and people. A real solution to that problem will therefore be ecologically, agriculturally, and culturally healthful.

Perhaps it is not until health is set down as the aim that we come in sight of the third kind of solution: that which causes a ramifying series of solutions—as when meat animals are fed on the farm where the feed is raised, and where the feed is raised to be fed to the animals that are on the farm. Even so rudimentary a description implies a concern for pattern, for quality, which necessarily complicates the concern for production. The farmer has put plants and animals into a relationship of mutual dependence, and must perforce be concerned for balance or symmetry, a reciprocating connection in the pattern of the farm that is biological, not industrial, and that involves solutions to problems of fertility, soil husbandry, economics, sanitation—the whole complex of problems whose proper solutions add up to *health:* the health of the soil, of plants and animals, of farm and farmer, of farm family and farm community, all involved in the same internested, interlocking pattern—or pattern of patterns.

A bad solution is bad, then, because it acts destructively upon the larger patterns in which it is contained. It acts destructively upon those patterns, most likely, because it is formed in ignorance or disregard of them. A bad solution solves for a single purpose or goal, such as increased production. And it is typical of such solutions that they achieve stupendous increases in production at exorbitant biological and social costs.

A good solution is good because it is in harmony with those larger patterns—and this harmony will, I think, be found to have the nature of analogy. A bad solution acts within the larger pattern the way a disease or addiction acts

within the body. A good solution acts within the larger pattern the way a healthy organ acts within the body. But it
must at once be understood that a healthy organ does
not—as the mechanistic or industrial mind would like to
say—"give" health to the body, is not exploited for the
body's health, but is *a part* of its health. The health of
organ and organism is the same, just as the health of organism and ecosystem is the same. And these structures of organ, organism, and ecosystem—as John Todd has so ably
understood—belong to a series of analogical integrities
that begins with the organelle and ends with the biosphere.

It would be next to useless, of course, to talk about the possibility of good solutions if none existed in proof and in
practice. A part of our work at *The New Farm* has been to
locate and understand those farmers whose work is competently responsive to the requirements of health. Representative of these farmers, and among them remarkable for the
thoroughness of his intelligence, is Earl F. Spencer, who has
a 250-acre dairy farm near Palatine Bridge, New York.

Before 1972, Earl Spencer was following a "conventional" plan which would build his herd to 120 cows. According to this plan, he would eventually buy all the grain he
fed, and he was already using as much as 30 tons per year
of commercial fertilizer. But in 1972, when he had increased his herd to 70 cows, wet weather reduced his harvest by about half. The choice was clear: he had either to
buy half his yearly feed supply, or sell half his herd.

He chose to sell half his herd—a very unconventional
choice, which in itself required a lot of independent intelligence. But character and intelligence of an even more respectable order were involved in the next step, which was
to understand that the initial decision implied a profound
change in the pattern of the farm and of his life and as-

sumptions as a farmer. With his herd now reduced by half, he saw that before the sale he had been overstocked, and had been abusing his land. On his 120 acres of tillable land, he had been growing 60 acres of corn and 60 of alfalfa. On most of his fields, he was growing corn three years in succession. The consequences of this he now saw as symptoms, and saw that they were serious: heavy dependence on purchased supplies, deteriorating soil structure, declining quantities of organic matter, increasing erosion, yield reductions despite continued large applications of fertilizer. In addition, because of his heavy feeding of concentrates, his cows were having serious digestive and other health problems.

He began to ask fundamental questions about the nature of the creatures and the land he was dealing with, and to ask if he could not bring about some sort of balance between their needs and his own. His conclusion was that "to be in balance with nature is to be successful." His farm, he says, had been going in a "dead run"; now he would slow it to a "walk."

From his crucial decision to reduce his herd, then, several other practical measures have followed:

1. A five-year plan (extended to eight years) to phase out entirely his use of purchased fertilizers.

2. A plan, involving construction of a concrete manure pit, to increase and improve his use of manure.

3. Better husbandry of cropland, more frequent rotation, better timing.

4. The gradual reduction of grain in the feed ration, and the concurrent increase of roughage—which has, to date, reduced the dependence on grain by half, from about 6000 pounds per cow to about 3000 pounds.

5. A breeding program which selects "for more efficient roughage conversion."

The most tangible results are that the costs of production

have been "dramatically" reduced, and that per cow production has increased by 1500 to 2000 pounds. But the health of the whole farm has improved. There is a moral satisfaction in this, of which Earl Spencer is fully aware. But he is also aware that the satisfaction is not *purely* moral, for the good results are also practical and economic: "We have half the animals we had before and are feeding half as much grain to those remaining, so we now need to plant corn only two years in a row. Less corn means less plowing, less fuel for growing and harvesting, and less wear on the most expensive equipment." Veterinary bills have been reduced also. And in 1981, if the schedule holds, he will buy no commercial fertilizer at all.

From the work of Earl Spencer and other exemplary farmers, and from the understanding of destructive farming practices, it is possible to devise a set of critical standards for agriculture. I am aware that the list of standards which follows must be to some extent provisional, but am nevertheless confident that it will work to distinguish between healthy and unhealthy farms, as well as between the oversimplified minds that solve problems for some X such as profit or quantity of production, and those minds, sufficiently complex, that solve for health or quality or coherence of pattern. To me, the validity of these standards seems inherent in their general applicability. They will serve the making of sewer systems or households as readily as they will serve the making of farms:

   1. A good solution accepts given limits, using so far as possible what is at hand. The farther-fetched the solution, the less it should be trusted. Granted that a farm can be too small, it is nevertheless true that enlarging scale is a deceptive solution; it solves one problem by acquiring another or several others.

2. A good solution accepts also the limitation of discipline. Agricultural problems should receive solutions that are agricultural, not technological or economic.

3. A good solution improves the balances, symmetries, or harmonies within a pattern—it is a qualitative solution—rather than enlarging or complicating some part of a pattern at the expense or in neglect of the rest.

4. A good solution solves more than one problem, and it does not make new problems. I am talking about health as opposed to almost any cure, coherence of pattern as opposed to almost any solution produced piecemeal or in isolation. The return of organic wastes to the soil may, at first glance, appear to be a good solution *per se*. But that is not invariably or necessarily true. It is true only if the wastes are returned to the right place at the right time in the pattern of the farm, if the waste does not contain toxic materials, if the quantity is not too great, and if not too much energy or money is expended in transporting it.

5. A good solution will satisfy a whole range of criteria; it will be good in all respects. A farm that has found correct agricultural solutions to its problems will be fertile, productive, healthful, conservative, beautiful, pleasant to live on. This standard obviously must be qualified to the extent that the pattern of the life of a farm will be adversely affected by distortions in any of the larger patterns that contain it. It is hard, for instance, for the economy of a farm to maintain its health in a national industrial economy in which farm earnings are apt to be low and expenses high. But it is apparently true, even in such an economy, that the farmers most apt to survive are those who do not go too far out of agriculture into either industry or banking—and who, moreover, live like farmers, not like businessmen. This seems especially true for the smaller farmers.

6. A good solution embodies a clear distinction between biological order and mechanical order, between farming

and industry. Farmers who fail to make this distinction are ideal customers of the equipment companies, but they often fail to understand that the real strength of a farm is in the soil.

7. Good solutions have wide margins, so that the failure of one solution does not imply the impossibility of another. Industrial agriculture tends to put its eggs into fewer and fewer baskets, and to make "going for broke" its only way of going. But to grow grain should not make it impossible to pasture livestock, and to have a lot of power should not make it impossible to use only a little.

8. A good solution always answers the question, How much is enough? Industrial solutions have always rested on the assumption that enough is all you can get. But that destroys agriculture, as it destroys nature and culture. The good health of a farm implies a limit of scale, because it implies a limit of attention, and because such a limit is invariably implied by any pattern. You destroy a square, for example, by enlarging one angle or lengthening one side. And in any sort of work there is a point past which more quantity necessarily implies less quality. In some kinds of industrial agriculture, such as cash grain farming, it is possible (to borrow an insight from Professor Timothy Taylor) to think of technology as a substitute for skill. But even in such farming that possibility is illusory; the illusion can be maintained only so long as the consequences can be ignored. The illusion is much shorter lived when animals are included in the farm pattern, because the husbandry of animals is so insistently a human skill. A healthy farm incorporates a pattern that a single human mind can comprehend, make, maintain, vary in response to circumstances, and pay steady attention to. That this limit is obviously variable from one farmer and farm to another does not mean that it does not exist.

9. A good solution should be cheap, and it should not

enrich one person by the distress or impoverishment of another. In agriculture, so-called "inputs" are, from a different point of view, outputs—*expenses*. In all things, I think, but especially in an agriculture struggling to survive in an industrial economy, any solution that calls for an expenditure to a manufacturer should be held in suspicion—not rejected necessarily, but *as a rule* mistrusted.

10. Good solutions exist only in proof, and are not to be expected from absentee owners or absentee experts. Problems must be solved in work and in place, with particular knowledge, fidelity, and care, by people who will suffer the consequences of their mistakes. There is no theoretical or ideal *practice*. Practical advice or direction from people who have no practice may have some value, but its value is questionable and is limited. The divisions of capital, management, and labor, characteristic of an industrial system, are therefore utterly alien to the health of farming—as they probably also are to the health of manufacturing. The good health of a farm depends on the farmer's mind; the good health of his mind has its dependence, and its proof, in physical work. The good farmer's mind and his body—his management and his labor—work together as intimately as his heart and his lungs. And the capital of a well-farmed farm by definition includes the farmer, mind and body both. Farmer and farm are one thing, an organism.

11. Once the farmer's mind, his body, and his farm are understood as a single organism, and once it is understood that the question of the endurance of this organism is a question about the sufficiency and integrity of a pattern, then the word *organic* can be usefully admitted into this series of standards. It is a word that I have been defining all along, though I have not used it. An organic farm, properly speaking, is not one that uses certain methods and substances and avoids others; it is a farm whose structure is formed in imitation of the structure of a natural system; it

has the integrity, the independence, and the benign dependence of an organism. Sir Albert Howard said that a good farm is an analogue of the forest which "manures itself." A farm that imports too much fertility, even as feed or manure, is in this sense as inorganic as a farm that exports too much or that imports chemical fertilizer.

12. The introduction of the term *organic* permits me to say more plainly and usefully some things that I have said or implied earlier. In an organism, what is good for one part is good for another. What is good for the mind is good for the body; what is good for the arm is good for the heart. We know that sometimes a part may be sacrificed for the whole; a life may be saved by the amputation of an arm. But we also know that such remedies are desperate, irreversible, and destructive; it is impossible to improve the body by amputation. And such remedies do not imply a safe logic. As *tendencies* they are fatal: you cannot save your arm by the sacrifice of your life.

Perhaps most of us who know local histories of agriculture know of fields that in hard times have been sacrificed to save a farm, and we know that though such a thing is possible it is dangerous. The danger is worse when topsoil is sacrificed for the sake of a crop. And if we understand the farm as an organism, we see that it is impossible to sacrifice the health of the soil to improve the health of plants, or to sacrifice the health of plants to improve the health of animals, or to sacrifice the health of animals to improve the health of people. In a biological pattern—as in the pattern of a community—the exploitive means and motives of industrial economics are immediately destructive and ultimately suicidal.

13. It is the nature of any organic pattern to be contained within a larger one. And so a good solution in one pattern preserves the integrity of the pattern that contains it. A good agricultural solution, for example, would not pollute

or erode a watershed. What is good for the water is good for the ground, what is good for the ground is good for plants, what is good for plants is good for animals, what is good for animals is good for people, what is good for people is good for the air, what is good for the air is good for the water. And vice versa.

14. But we must not forget that those human solutions that we may call organic are not natural. We are talking about organic *artifacts,* organic only by imitation or analogy. Our ability to make such artifacts depends on virtues that are specifically human: accurate memory, observation, insight, imagination, inventiveness, reverence, devotion, fidelity, restraint. Restraint—for us, now—above all: the ability to accept and live within limits; to resist changes that are merely novel or fashionable; to resist greed and pride; to resist the temptation to "solve" problems by ignoring them, accepting them as "trade-offs," or bequeathing them to posterity. A good solution, then, must be in harmony with good character, cultural value, and moral law.

III

10

The Economics
of Subsistence
(1980)

One of the ideas most ruinous to the small farm has been that the farmer "could not afford" to produce his own food; the time and acreage required for the family's subsistence could be better used for market production. And so arrived that most curious manifestation of agricultural progress: farm families buying meat, vegetables, milk, and eggs at the supermarket "just like city people."

I don't believe, myself, that this "rule" was ever true. It was invented by "agribusiness" for the benefit of "agribusiness." It was not meant to help farmers, and it has not helped them. But even supposing that it may once have been true, it is true no longer.

According to the Fall 1979 *Family Economics Review*, the annual cost of food for a family of four (a couple 20–54 years old, and two children aged 6–8 and 9–11) varied at that time from $2,493.60 to $4,887.60. These figures were based on the assumption "that food for all meals and snacks is purchased at the store and prepared at home." And they were computed at cost levels of June 1979. The lower figure was based on a "thrifty plan," the lowest of four plans; the highest figure was based on a "liberal plan," the highest of the four.

There is necessarily some conjecture in deciding where a home-grown subsistence would fit into this scale of plans. Because both quality and variety of home-grown food tend to be optimal, it seems reasonable to conclude that the closest resemblance would be to the "liberal plan." But if you figure the value, not according to what is produced, but strictly according to what is saved, then you must probably say that you save the value of the "thrifty plan." And because of the unlikelihood that any family will produce *everything*, you must deduct something for the staples, etc., that must be purchased. Let us say, then, that the value of a home-grown subsistence, at the prices of June 1979, would be between about $2,000 and about $4,000. Even with inflation, the least of these figures is still a lot of money. It would pay the annual interest on quite a large loan.

What is required for the production of this $2,000 to $4,000 worth of subsistence? Well, first of all, a vegetable garden, perhaps with some small fruits, perhaps with a few orchard trees. This can be accommodated easily on half an acre of ground. If you garden organically, there will be no cost for fertilizer, and only a minimal cost for some insecticide such as rotenone. The major outlay other than labor will be for seeds, and this can be reduced by saving at least some of your own.

Next in importance, I think, is a milk cow — or, more exactly, what you might call a milk cow economy: the cow for milk, cream, and butter; her calf for beef; a meat hog to consume the surplus of skimmed milk (also kitchen scraps, residues from the garden, etc.). The economic value of one family milk cow, used in this way, is most impressive. (I say *one* cow, because the value of the second cow will be a great deal less. Given the nature of the larger economy, the value of the produce you eat is much greater than that of the produce you sell.)

For example, our own family cow is a small Jersey, who gives very rich milk in return for comparatively little feed. Her real worth to us is a little difficult to determine because, if you have a cow, you are apt to use more dairy products than you would if you did not have one. But if (as I estimate) my family uses three quarts of milk a day, and two pounds of butter and a quart of cream a week, then our cow is supplying us products worth something like $10 per week. If we milk her forty-four weeks a year her gross annual "earning" is $440. But in the past year, in addition to keeping us abundantly supplied with milk, cream, butter, etc., our cow produced a slaughter calf that weighed, I would say, 400—450 lbs., and our surplus of skimmed milk made a significant part of the diet of a meat hog that finished out at about 400 lbs.

In addition to pasture, our cow ate perhaps seventy-five bales (2500 – 3000 lbs.) of hay and perhaps the equivalent of twenty bushels of ear corn. I do not think the money value of her feed could possibly have exceeded $175.

I want to emphasize that the figures in the two preceding paragraphs are estimates. To get exact figures on the performance of a single animal, where numerous other animals are also being fed, would require more record keeping than I expect ever to have time to do. For the same reason, I have not even tried to estimate the value of milk in the diets of the calf and the hog. But I have tried to be conservative in determining all the amounts involved, and I think that my figures are reasonable.

On the basis of the dairy budget alone, our cow has put us ahead this year by something like $265. In addition to that, her contribution to the making of our beef and pork must be said to have been substantial (certainly well above whatever her calf would have brought as a veal.) And, of course, there is the value of her manure. But that is not all, for after her last year's calf was weaned, our cow raised two

Holstein "bucket calves" which cost $120 and brought $377.26, leaving $257.26, which I believe easily paid both for the cow's feed, and for the grain and hay fed to the three calves and the hog.

This cow grazed two small fields in rotation. I do not know the combined acreage, but would guess it to be well under two acres—and of common land. An acre of good pasture would have been enough. To produce her hay and grain would have taken about one-half acre of fertile ground.

Cow and garden, then, would require at most two acres of good ground. Setting the worth of subsistence at the lowest available figure—that of the "thrifty plan"—those two acres would be producing an annual net income of around $2,000, a figure that we do not fully appreciate until we realize that no more than half of this acreage is in row crops. I do not know how else land could be used so conservatively for so large a return.

Another important part of subsistence is heating fuel. If you own a woodlot and heat with wood, you will accomplish a saving that will be a considerable "income" from your farm. The amount you save will vary, of course, with climate, the size of your house, etc., but I believe that a saving of $1,000 would not be unlikely or unusual. That would raise the value of your subsistence to at least $3,000. If you are eating on the "liberal plan" and heating with your own wood, then your subsistence "income" amounts to $5,000.

If you live on a "marginal" thirty-acre farm and take nothing but your food and fuel from it, then your farm will be producing at an annual rate of from $100 to $166.66 per acre—and even $100 is a far better per-acre income than the "agribusinessmen" make by growing corn.

Of course, when the experts started telling farmers that they couldn't afford to produce their subsistence, they were talking about "economies of scale" and the cost of "inputs," the chief of which being labor. If you are talking about a farmer whose "operation" consists exclusively of 1000 acres of corn and soybeans, then I suppose you could demonstrate (on paper, anyhow) that he "could not afford" to raise a garden or milk a cow. A farmer with that kind of "operation" is, of course, a machine part, and if he wants to stay friends with the bank, he has to keep himself screwed to his tractor.

But if you are talking about a farm instead of an "operation," I think the argument changes radically. By *farm* I mean a place that is used diversely and conservingly, that grows animals as well as plants, that is of a size appropriate to the needs and the available energy of a family. On such a farm, where you put up hay and harvest grain for feed anyway, and where you have chores to do anyway, I don't think you are going to take much notice of the time and energy you invest in a milk cow, her calf, and a hog or two. If you are going to be at the barn for other chores, you may expect to "lose" only a few minutes a day. And if you haven't missed your calling in becoming a farmer, you may not think of caring for stock as a "loss."

How much time does a garden take? There is probably no sure way to tell. There are too many variables. Doing work at the right time almost always saves time. A push plow is faster than a hoe; a rototiller is faster than a push plow; a horse-drawn single-row cultivator is faster (and a better weeder) than a rototiller. If you want your garden to look picture perfect you will have to spend more time at work than someone who doesn't mind a few weeds. If you think of gardening as a kind of agony, then perhaps the investment of work may be too great, and you should not do it. But in that case, should you be farming? If you think of gardening as a pleasure, then you may not think of your work as an "investment" at all; the whole enterprise (less seed) will be profit. Or you may think of a garden as an economic necessity, which well repays the time invested in it. But that raises another question.

What is your time worth? Though it is often asked, I do not think this question is answerable. It is the same as asking what your life is worth. And I can give it only the same nonanswer: it is worth whatever it means. The idea that you cannot afford to raise a garden is based on the assumption that it means money, that if you are not receiving the top dollar for every minute of your life, you are suffering a "loss"—a doctrine that would not only put an end to gardens, but soon drive us all to theft or suicide.

A better question is this: What would you—and your children—be doing with the time "saved" by *not* producing your subsistence? Different families will answer that question in different ways. But if the answer is that you would be doing something expensive away from home, or planting a lot of extra acreage in corn or soybeans to pay for groceries and heating oil, or taking an outside job to buy food, or if you don't know what you would do with the time, then it seems to me that you would do well to look carefully into the economics of subsistence.

11

Family Work
(1980)

For those of us who have wished to raise our food and our children at home, it is easy enough to state the ideal. Growing our own food, unlike buying it, is a complex activity, and it affects deeply the shape and value of our lives. We like the thought that the outdoor work that improves our health should produce food of excellent quality that, in turn, also improves and safeguards our health. We like no less the thought that the home production of food can improve the quality of family life. Not only do we intend to give our children better food than we can buy for them at the store, or than they will buy for themselves from vending machines or burger joints, we also know that growing and preparing food at home can provide family work—work for everybody. And by thus elaborating household chores and obligations, we hope to strengthen the bonds of interest, loyalty, affection, and cooperation that keep families together.

Forty years ago, for most of our people, whether they lived in the country or in town, this was less an ideal than a necessity, enforced both by tradition and by need. As is often so, it was only after family life and family work became (allegedly) unnecessary that we began to think of them as "ideals."

As ideals, they are threatened; as they have become (even allegedly) unnecessary, they have become by the same token less possible. I do not mean to imply that *I* think the ideal is any less valuable than it ever was, or that it is—in reality, in the long run—less necessary. Nor do I think that less possible means impossible.

I do think that the ideal is more difficult now than it was. We are trying to uphold it now mainly by will, without much help from necessity, and with no help at all from custom or public value. For most people now do seem to think that family life and family work are unnecessary, and this thought has been institutionalized in our economy and in our public values. Never before has private life been so preyed upon by public life. How can we preserve family life—if by that we mean, as I think we must, *home* life— when our attention is so forcibly drawn away from home?

We know the causes well enough.

Automobiles and several decades of supposedly cheap fuel have put longer and longer distances between home and work, household and daily needs.

TV and other media have learned to suggest with increasing subtlety and callousness—especially, and most wickedly, to children—that it is better to consume than to produce, to buy than to grow or to make, to "go out" than to stay home. If you have a TV, your children will be subjected almost from the cradle to an overwhelming insinuation that all worth experiencing is somewhere else and that all worth having must be bought. The purpose is blatantly to supplant the joy and beauty of health with cosmetics, clothes, cars, and ready-made desserts. There is clearly too narrow a limit on how much money can be made from health, but the profitability of disease—especially disease of spirit or character—has so far, for profiteers, no visible limit.

Another cause, and one that seems particularly regrettable, is public education. The idea that the public should be educated is altogether salutary, and since we insist on making this education compulsory we ought, in reason, to reconcile ourselves to the likelihood that it will be mainly poor. I am not nearly so much concerned about its quality as I am about its *length*. My impression is that the chief, if unadmitted, purpose of the school system is to keep children away from home as much as possible. Parents want their children kept out of their hair; education is merely a by-product, not overly prized. In many places, thanks to school consolidation, two hours or more of travel time have been added to the school day. For my own children the regular school day from the first grade—counting from the time they went to catch the bus until they came home—was nine hours. An extracurricular activity would lengthen the day to eleven hours or more. This is not education, but a form of incarceration. Why should anyone be surprised if, under these circumstances, children should become "disruptive" or even "ineducable"?

If public education is to have any meaning or value at all, then public education *must* be supplemented by home education. I know this from my own experience as a college teacher. What can you teach a student whose entire education has been public, whose daily family life for twenty years has consisted of four or five hours of TV, who has never read a book for pleasure or even *seen* a book so read; whose only work has been schoolwork, who has never learned to perform any essential task? Not much, so far as I could tell.

We can see clearly enough at least a couple of solutions. We can get rid of the television set. As soon as we see that the TV cord is a vacuum line, pumping life and meaning out of the household, we can unplug it. What a grand and neglected privilege it is to be shed of the glibness, the gleeful idiocy, the idiotic gravity, the unctuous or lubricious greed of those public faces and voices!

And we can try to make our homes centers of attention and interest. Getting rid of the TV, we understand, is not just a practical act, but also a symbolical one: we thus turn our backs on the invitation to consume; we shut out the racket of consumption. The ensuing silence is an invitation to our homes, to our own places and lives, to come into being. And we begin to recognize a truth disguised or denied by TV and all that it speaks and stands for: no life and no place is destitute; all have possibilities of productivity and pleasure, rest and work, solitude and conviviality that belong particularly to themselves. These possibilities exist everywhere, in the country or in the city, it makes no difference. All that is necessary is the time and the inner quietness to look for them, the sense to recognize them, and the grace to welcome them. They are now most often lived out in home gardens and kitchens, libraries, and workrooms. But they are beginning to be worked out, too, in little parks, in vacant lots, in neighborhood streets. Where we live is also a place where our interest and our effort can be. But they can't be there by the means and modes of consumption. If we consume nothing but what we buy, we are living in "the economy," in "television land," not at home. It is productivity that rights the balance, and brings us home. Any way at all of joining and using the air and light and weather of your own place— even if it is only a window box, even if it is only an opened window—is a making and a having that you cannot get from TV or government or school.

That local productivity, however small, is a gift. If we are parents we cannot help but see it as a gift to our children—and the *best* of gifts. How will it be received?

Well, not ideally. Sometimes it will be received gratefully enough. But sometimes indifferently, and sometimes resentfully.

According to my observation, one of the likeliest results of a wholesome diet of home-raised, home-cooked food is a heightened relish for cokes and hot dogs. And if you "deprive" your children of TV at home, they are going to watch it with something like rapture away from home. And obligations, jobs, and chores at home will almost certainly cause your child to wish, sometimes at least, to be somewhere else, watching TV.

Because, of course, parents are not the only ones raising their children. They are being raised also by their schools and by their friends and by the parents of their friends. Some of this outside raising is good, some is not. It is, anyhow, unavoidable.

What this means, I think, is about what it has always meant. Children, no matter how nurtured at home, must be risked to the world. And parenthood is not an exact science, but a vexed privilege and a blessed trial, absolutely necessary and not altogether possible.

If your children spurn your healthful meals in favor of those concocted by some reincarnation of Col. Sanders, Long John Silver, or the Royal Family of Burger; if they flee from books to a friend's house to watch TV, if your old-fashioned notions and ways embarrass them in front of their friends—does that mean you are a failure?

It may. And what parent has not considered that possibility? I know, at least, that I have considered it—and have wailed and gnashed my teeth, found fault, laid blame, preached and ranted. In weaker moments, I have even blamed myself.

But I have thought, too, that the term of human judgment is longer than parenthood, that the upbringing we give our children is not just for their childhood but for all their lives. And it is surely the *duty* of the older generation to be embarrassingly old-fashioned, for the claims of the "newness" of any younger generation are mostly frivolous. The young are born to the human condition more than to their time, and they face mainly the same trials and obligations as their elders have faced.

The real failure is to give in. If we make our house a household instead of a motel, provide healthy nourishment for mind and body, enforce moral distinctions and restraints, teach essential skills and disciplines and require their use, there is no certainty that we are providing our children a "better life" that they will embrace wholeheartedly during childhood. But we are providing them a choice that they may make intelligently as adults.

12

The Reactor
and the Garden
(1979)

On June 3, 1979, I took part in an act of nonviolent civil disobedience at the site of a nuclear power plant being built at Marble Hill, near Madison, Indiana. At about noon that day, eighty-nine of us crossed a wire fence onto the power company's land, were arrested, and duly charged with criminal trespass.

As crimes go, ours was tame almost to the point of boredom. We acted under a well-understood commitment to do no violence and damage no property. The Jefferson County sheriff knew well in advance and pretty exactly what we planned to do. Our trespass was peaceable and orderly. We were politely arrested by the sheriff and his deputies, who acted, as far as I saw, with exemplary kindness. And this nearly eventless event ended in anticlimax: the prosecutor chose to press charges against only one of the eighty-nine who were arrested, and that one was never brought to trial.

And yet, for all its tameness, it was not a lighthearted event. Few of us, I think, found it easy to decide to break the law of the land. For me it was difficult for another reason as well: I do not like public protests or crowd actions of any kind; I dislike and distrust the slogans and the jargon that invariably stick like bubble gum to any kind of "movement."

Why did I do it?

For several years, along with a good many other people, I have been concerned about the proliferation of power plants in the Ohio River Valley, where at present more than sixty plants are either working, under construction, or planned. Air pollution from existing coal-fired plants in the valley is already said to be the worst in the country. And the new plants are being constructed or planned without any evident consideration of the possibility of limiting or moderating the consumption of electricity. The people of this area, then, are expected to sacrifice their health — among other things — to underwrite the fantasy of "unlimited economic growth." This is a decision not made by them — but, rather, made *for* them by the power companies in collaboration with various agencies of government.

The coal-fired plants would be bad enough by themselves. But, in addition, some power companies have decided that nuclear power is the best answer to "the energy problem," and two nuclear power plants are now under construction in this part of the Ohio Valley. The arguments in their favor are not good, but they are backed nevertheless by a great deal of money and political power. For example, our local rural electric co-op publishes a magazine which constantly editorializes in favor of nuclear power. The rate payers are thus, in effect, being taxed to promote an energy policy that many of them consider objectionable and dangerous.

Power plants in the Ohio Valley raise another serious problem, this one political. The Ohio River is a state boundary. A power plant on the north side of the river in Indiana will obviously have an effect in Kentucky. But though a plant will necessarily affect at least two states, it is planned and permitted only in one. The people of one state thus become subject to a decision made in another state, in which they are without representation. And so in the behavior of big technology and corporate power, we can recognize again an exploitive colonialism similar to that of George III.

Like the majority of people, I am unable to deal competently with the technical aspects of nuclear power and its dangers. My worries are based on several facts available to any reader of a newspaper:

1. Nuclear power is extremely dangerous. For this, the elaborate safety devices and backup systems of the plants themselves are evidence enough. Radioactive wastes, moreover, remain dangerous for many of thousands of years, and there is apparently no foreseeable safe way to dispose of them.

2. Dangerous accidents do happen in nuclear power plants. Officials and experts claim that accidents can be foreseen and prevented, but accidents are surprises by definition. If they are foreseen they do not happen.

3. Nuclear experts and plant employees do not always act competently in dealing with these accidents. Nuclear power requires people to act with *perfect* competence if it is to be used safely. But people in nuclear power plants are just as likely to blunder or panic or miscalculate as people anywhere else.

4. Public officials do not always act responsibly. Sometimes they deliberately falsify, distort, or withhold information essential to the public's health or safety.

If I had doubts about any of this, they were removed forever by the accident at Three Mile Island. And if I had any lingering faith that the government would prove a trustworthy guardian of public safety, that was removed by the recent hearings on the atomic bomb tests of the 1950s—which have revealed that the government assured the people living near the explosions that there would be no danger from radiation, when in fact it knew that the danger would be great.

And so when I climbed the fence at Marble Hill, I considered that I was casting a vote that I had been given no better opportunity to cast. I was voting no. And I was voting no confidence. Marble Hill is only about twenty miles upwind from my house. As a father, a neighbor, and a citizen, I had begun to look on the risk of going to jail as trivial in comparison to the risks of living so near a nuclear power plant.

But even though I took part wholeheartedly in the June 3 protest, I am far from believing that such public acts are equal to their purpose, or that they ever will be. They are necessary, but they are not enough, and they subject the minds of their participants to certain dangers.

Any effort that focuses on one problem encourages oversimplification. It is easy to drift into the belief that once the nuclear power problem—or the energy problem, or the pollution problem—is solved everything will be all right. It will not, of course. For all these separate problems are merely aspects of the human problem, which never has been satisfactorily solved, and which would provide every one of us a lifetime agenda of work and worry even if *all* the bedeviling problems of twentieth century technology were solved today.

An even greater danger is that of moral oversimplification, or self-righteousness. Protests, demonstrations, and other forms of "movement" behavior tend to divide people into the ancient categories of "us" and "them." In the midst of the hard work and the risks of opposing what "we" see as a public danger, it is easy to assume that if only "they" were as clear-eyed, alert, virtuous, and brave as "we" are, our problems would soon be solved. This notion, too, is patently false. In the argument over nuclear power—as in most public arguments—the division between "us" and "them" does not really exist. In our efforts to correct the way things are, we are almost always, almost inevitably, opposing what is wrong with ourselves. If we do not see that, then I think we won't find any of the solutions we are looking for.

For example, I believe that most people who took part in the June 3 demonstration at Marble Hill got there in an automobile. I did, and I could hardly have got there any other way. Thus the demonstration, while it pushed for a solution to one aspect of the energy problem, was itself another aspect of that problem.

And I would be much surprised to learn that most of us did not return home to houses furnished with electric light switches, which we flipped on more or less thoughtlessly, not worrying overmuch about the watersheds that are being degraded or destroyed by strip mines to produce the coal to run the power plants to make the electricity that burns in our light bulbs. I know, anyhow, that I often flip on my own light switches without any such worries.

Nearly all of us are sponsoring or helping to cause the ills we would like to cure. Nearly all of us have what I can only call cheap-energy minds; we continue to assume, or to act as if we assume, that it does not matter how much energy we use.

I do not mean to imply that I know how to solve the problems of the automobile or of the wasteful modern household. Those problems are enormously difficult, and their difficulty suggests their extreme urgency and importance. But I am fairly certain that they won't be solved simply by public protests. The roots of the problems are private or personal, and the roots of the solutions will be private or personal too. Public protests are incomplete actions; they speak to the problem, not to the solution.

Protests are incomplete, I think, because they are by definition negative. You cannot protest *for* anything. The positive thing that protest is supposed to do is "raise consciousness," but it can raise consciousness only to the level of protest. So far as protest itself is concerned, the raised consciousness is on its own. It appears to be possible to "raise" your consciousness without changing it—and so to keep protesting forever.

If you have to be negative, there are better negative things to do. You can quit doing something you know to be destructive. It might, for instance, be possible to take a pledge that you will no longer use electricity or petroleum to entertain yourself. My own notion of an ideal negative action is to get rid of your television set. (It is cheating to get rid of it by selling it or giving it away. You should get rid of it by carefully disassembling it with a heavy blunt instrument. Would you try to get rid of any other brain disease by selling it or giving it away?)

But such actions are not really negative. When you get rid of something undesirable you are extending an invitation to something desirable. If it is true that nature abhors a vacuum, there is no need to fear. Wherever you make an opening, it will be filled. When you get rid of petroleum-powered or electronic entertainment you are inviting a renewal of that structure of conversation, work, and play that used to be known as "home life." You are inviting such gentle and instructive pleasures as walking and reading.

Or it may be possible for some people to walk or ride a bicycle to work—and so to consider doing without a car altogether. Or there may be some kind of motor-powered tool that can be done without. Or perhaps it will prove economical or pleasing to change from fossil fuel heat to a solar collector or a wood stove.*

There is, then, a kind of negative action that cannot re-

*But the use of wood stoves without proper maintenance of wood lots is only another form of mining. It makes trees an exhaustible resource.

main negative. To give up some things is to create problems, which immediately call for solutions—and so the negative action completes itself in an action that is positive. But some actions are probably more complete than others, and the more complete the action, the more effective it is as a protest.

What, then, is a complete action? It is, I think, an action which one takes on one's own behalf, which is particular and complex, real not symbolic, which one can both accomplish on one's own and take full responsibility for. There are perhaps many such actions, but certainly among them is any sort of home production. And of the kinds of home production, the one most possible for most people is gardening.

Some people will object at this point that it belittles the idea of gardening to think of it as an act of opposition or protest. I agree. That is exactly my point. Gardening—or the best kind of gardening—is a *complete* action. It is so effective a protest because it is so much more than a protest.

The best kind of gardening is a form of home production capable of a considerable independence of outside sources. It will, then, be "organic" gardening. One of the most pleasing aspects of this way of gardening is its independence. For fertility, plant protection, etc., it relies as far as possible on resources in the locality and in the gardener's mind. Independence can be further enlarged by saving seed and starting your own seedlings. To work at ways of cutting down the use of petroleum products and gasoline engines in the garden is at once to increase independence and to work directly at a real (that is, a permanent) solution to the energy problem.

A garden gives interest a place, and it proves one's place interesting and worthy of interest. It works directly against the feeling—the source of a lot of our "environmental" troubles—that in order to be diverted or entertained, or to "make life interesting," it is necessary to draw upon some distant resource—turn on the TV or take a trip.

One of the most important local resources that a garden makes available for use is the gardener's own body. At a time when the national economy is largely based on buying and selling substitutes for common bodily energies and functions, a garden restores the body to its usefulness—a victory for our species. It may take a bit of effort to realize that perhaps the most characteristic modern "achievement" is the obsolescence of the human body. Jogging and other forms of artificial exercise do not restore the usefulness of the body, but are simply ways of assenting to its uselessness; the body is a diverting pet, like one's Chihuahua, and must be taken out for air and exercise. A garden gives the body the dignity of working in its own support. It is a way of rejoining the human race.

One of the common assumptions, leading to the obsolescence of the body, is that physical work is degrading. That is true if the body is used as a slave or a machine—if, in other words, it is misused. But working in one's own garden does not misuse the body, nor does it dull or "brutalize" the mind. The work of gardening is not "drudgery," but is the finest sort of challenge to intelligence. Gardening is not a discipline that can be learned once for all, but keeps presenting problems that must be directly dealt with. It is, in addition, an agricultural and ecological education, and that sort of education corrects the cheap-energy mind.

A garden is the most direct way to recapture the issue of health, and to make it a private instead of a governmental responsibility. In this, as in several other ways I have mentioned, gardening has a power that is political and even democratic. And it is a political power that can be applied constantly, whereas one can only vote or demonstrate occasionally.

Finally, because it makes backyards (or front yards or vacant lots) productive, gardening speaks powerfully of the abundance of the world. It does so by increasing and enhancing abundance, and by demonstrating that abundance, given moderation and responsible use, is limitless. We learn from our gardens to deal with the most urgent question of the time: How much is enough? We don't soup our gardens up with chemicals because our goal is *enough,* and we know that *enough* requires a modest, moderate, conserving technology.

Atomic reactors and other big-technological solutions, on the other hand, convey an overwhelming suggestion of the poverty of the world and the scarcity of goods. That is because their actuating principle is excessive consumption. They obscure and destroy the vital distinction between abundance and extravagance. The ideal of "limitless economic growth" is based on the obsessive and fearful conviction that more is always needed. The growth is maintained by the consumers' panic-stricken suspicion, since they always want more, that they will never have enough.

Enough is everlasting. Too much, despite all the ballyhoo about "limitless growth," is temporary. And big-technological solutions are temporary: the lifetime of a nuclear power plant is thirty years! A garden, given the right methods and the right care, will last as long as the world.

A garden, of course, is not always as comfortable as Kroger's. If you grow a garden you are going to shed some sweat, and you are going to spend some time bent over; you will experience some aches and pains. But it is in the willingness to accept this discomfort that we strike the most telling blow against the power plants and what they represent. We have gained a great deal of comfort and convenience by our dependence on various public utilities and government agencies. But it is obviously not possible to become dependent without losing independence—and freedom too. Or to put it another way, we cannot be free from discomfort without becoming subject to the whims and abuses of centralized power, and to any number of serious threats to our health. We cannot hope to recover our freedom from such perils without discomfort.

Someone is sure to ask how I can suppose that a garden, "whose action is no stronger than a flower," can compete with a nuclear reactor. Well, I am not supposing that exactly. As I said, I think the protests and demonstrations are necessary. I think that jail may be the freest place when you *have no choice* but to breathe poison or die of cancer. But it is futile to attempt to correct a public wrong without correcting the sources of that wrong in yourself.

At the same time, I think it may be too easy to underestimate the power of a garden. A nuclear reactor is a proposed "solution" to "the energy problem." But like all big-technological "solutions," this one "solves" a single problem by causing many. The problems of what to do with radioactive wastes and with decommissioned nuclear plants, for example, have not yet been solved; and we can confidently predict that the "solutions," when they come, will cause yet other serious problems that will come as "surprises" to the officials and the experts. In that way, big technology works perpetually against itself. That is the limit of "unlimited economic growth."

A garden, on the other hand, is a solution that leads to other solutions. It is a part of the limitless pattern of good health and good sense.

13

A Good Scythe
(1979)

When we moved to our little farm in the Kentucky River Valley in 1965, we came with a lot of assumptions that we have abandoned or changed in response to the demands of place and time. We assumed, for example, that there would be good motor-powered solutions for all of our practical problems.

One of the biggest problems from the beginning was that our place was mostly on a hillside and included a good deal of ground near the house and along the road that was too steep to mow with a lawn mower. Also, we were using some electric fence, which needed to be mowed out once or twice a year.

When I saw that Sears Roebuck sold a "power scythe," it seemed the ideal solution, and I bought one. I don't remember what I paid for it, but it was expensive, considering the relatively small amount of work I needed it for. It consisted of a one-cylinder gasoline engine mounted on a frame with a handlebar, a long metal tube enclosing a flexible drive shaft, and a rotary blade. To use it, you hung it from your shoulder by a web strap, and swept the whirling blade over the ground at the desired height.

It did a fairly good job of mowing, cutting the grass and

weeds off clean and close to the ground. An added advantage was that it readily whacked off small bushes and tree sprouts. But this solution to the mowing problem involved a whole package of new problems:

1. The power scythe was heavy.

2. It was clumsy to use, and it got clumsier as the ground got steeper and rougher. The tool that was supposed to solve the problem of steep ground worked best on level ground.

3. It was dangerous. As long as the scythe was attached to you by the shoulder strap, you weren't likely to fall onto that naked blade. But it *was* a naked blade, and it did create a constant threat of flying rock chips, pieces of glass, etc.

4. It enveloped you in noise, and in the smudge and stench of exhaust fumes.

5. In rank growth, the blade tended to choke—in which case you had to kill the engine in a hurry or it would twist the drive shaft in two.

6. Like a lot of small gas engines not regularly used, this one was temperamental and undependable. And dependence on an engine that won't run is a plague and a curse.

When I review my own history, I am always amazed at how slow I have been to see the obvious. I don't remember how long I used that "labor-saving" power scythe before I finally donated it to help enlighten one of my friends—but it was too long. Nor do I remember all the stages of my own enlightenment.

The turning point, anyhow, was the day when Harlan Hubbard showed me an old-fashioned, human-powered scythe that was clearly the best that I had ever seen. It was light, comfortable to hold and handle. The blade was very sharp, angled and curved precisely to the path of its stroke. There was an intelligence and refinement in its design that made it a pleasure to handle and look at and think about.

I asked where I could get one, and Harlan gave me an address: The Marugg Company, Tracy City, Tennessee 37387.

I wrote for a price list and promptly received a sheet exhibiting the stock in trade of the Marugg Company: grass scythes, bush scythes, snaths, sickles, hoes, stock bells, carrying yokes, whetstones, and the hammers and anvils used in beating out the "dangle" cutting edge that is an essential feature of the grass scythes.

In due time I became the owner of a grass scythe, hammer and anvil, and whetstone. Learning to use the hammer and anvil properly (the Marugg Company provides a sheet of instructions) takes some effort and some considering. And so does learning to use the scythe. It is essential to hold the point so that it won't dig into the ground, for instance; and you must learn to swing so that you slice rather than hack.

Once these fundamentals are mastered, the Marugg grass scythe proves itself an excellent tool. It is the most satisfying hand tool that I have ever used. In tough grass it cuts a little less uniformly than the power scythe. In all other ways, in my opinion it is a better tool:

1. It is light.

2. It handles gracefully and comfortably even on steep ground.

3. It is far less dangerous than the power scythe.

4. It is quiet and makes no fumes.

5. It is much more adaptable to conditions than the power scythe: in ranker growth, narrow the cut and shorten the stroke.

6. It always starts—provided the user will start. Aside from reasonable skill and care in use, there are no maintenance problems.

7. It requires no fuel or oil. It runs on what you ate for breakfast.

8. It is at least as fast as the power scythe. Where the cutting is either light or extra heavy, it can be appreciably faster.

9. It is far cheaper than the power scythe, both to buy and to use.

Since I bought my power scythe, a new version has come on the market, using a short length of nylon string in place of the metal blade. It is undoubtedly safer. But I believe the other drawbacks remain. Though I have not used one of these, I have observed them in use, and they appear to me to be slower than the metal-bladed power scythe, and less effective on large-stemmed plants.

I have noticed two further differences between the power scythe and the Marugg scythe that are not so practical as those listed above, but which I think are just as significant. The first is that I never took the least pleasure in using the power scythe, whereas in using the Marugg scythe, whatever the weather and however difficult the cutting, I always work with the pleasure that one invariably gets from using a good tool. And because it is not motor driven and is quiet and odorless, the Marugg scythe also allows the pleasure of awareness of what is going on around you as you work.

The other difference is between kinds of weariness. Using the Marugg scythe causes the simple bodily weariness that comes with exertion. This is a kind of weariness that, when not extreme, can in itself be one of the pleasures of work. The power scythe, on the other hand, adds to the weariness of exertion the unpleasant and destructive weariness of strain. This is partly because, in addition to carrying and handling it, your attention is necessarily clenched to it; if you are to use it effectively and safely, you *must* not look away. And partly it is because the power scythe, like all motor-driven tools, imposes patterns of endurance that are alien to the body. As long as the motor is running there is a pressure to keep going. You don't stop to consider or rest or look around. You keep on until the motor stops or the job is finished or you have some kind of trouble. (This explains why the tractor soon evolved headlights, and farmers began to do daywork at night.)

These differences have come to have, for me, the force of a parable. Once you have mastered the Marugg scythe, what an absurd thing it makes of the power scythe! What possible sense can there be in carrying a heavy weight on your shoulder in order to reduce by a very little the use of your arms? Or to use quite a lot of money as a substitute for a little skill?

The power scythe—and it is far from being an isolated or unusual example—is *not* a labor saver or a shortcut. It is a labor maker (you have to work to pay for it as well as to use it) and a long cut. Apologists for such expensive technological solutions love to say that "you can't turn back the clock." But when it makes perfect sense to do so—as when the clock is wrong—of *course* you can!

14

Looking Ahead
(1978)

The university intellectuals are increasingly preoccupied with the future. They are not especially interested in *preparing* for the future—which is something that people do by behaving considerately, moderately, conservingly, and decently in the present—but in *predicting* the future, saying now what will happen then. But one of the fundamental truths of human experience is that we can never be sure what will happen in the next minute, much less in the next century. So what are the reasons for all these botherings of the future by the so-called "futurologists?" I will suggest several.

First, "futurology" is a new academic profession; as such, it provides a lot of new "job openings." The universities, in order to give their graduate faculties work appropriate to their dignity have turned out too many doctors of philosophy. One way to deal with such a surplus is to featherbed: if the market for these people's brains is oversupplied in the present, then put them to work in the future. Thus is born, or contrived, the "futurologist." He escapes present un- or underemployment by going to work in the future *now*.

Second, "futurology" has suddenly become a very lucrative and glamorous specialty. "Futurologists" are much in demand as conferees and consultants. What they say or write is almost certain to attract public attention. And they give academic prestige to the purposes and ambitions of the corporations. Their "projections" and "worlds of the future" almost invariably rest on the assumption that society cannot help but become more centralized and mechanized, and that people will become more dependent on products that they cannot produce for themselves.

Third, the future is the best of all possible settings for the airy work of academic theorists—simply because neither nature nor human nature has yet taken place there. If you build a castle in the air *now*, people will notice that it does not exist, and you will be accused of pipedreaming; people will think you are crazy. But if you build it in the future, which does not exist, you can call it a "logical projection," your colleagues will talk learnedly about it, you can hope to get a promotion, a salary increase, and to earn large fees as a lecturer and consultant.

These "logical projections" remind me of what farmers sometimes call "winter crops." A winter crop is the crop that a farmer grows in his mind while he sits by the stove in the winter. They are always perfect crops. They are perfect because no sweat has been shed in them, and they are safe from pests, human frailty, and bad weather. Summer crops are another matter. "Futurology" is a way to secure a professional salary for a grower of winter crops; a "futurologist" is a person who *never* needs to worry about growing a crop in the summertime. "That ain't what I call a job," says a neighbor of mine. "That's what I call a position."

The most recent "logical projection" that I have seen is the work of eleven engineering professors at Purdue University. This one proposes to tell us what American life will be like at the beginning of the twenty-first century, and I venture to say that nobody has ever pipedreamed a more dismal "logical projection." The account I read offers a glimpse of the daily life in 2001 of "the fictitious Niray family, living in the imaginary Midland City, U.S.A." A few samples of the text will be enough to show what a perfect "world of the future" this is—for machines.

The hero of this fiction, Dave Niray, breakfasted on a "cylinder of Nutri-Juice;" in 2001 nobody cooked at home but a few eccentrics: gourmets and old-fashioned people. (For some unexplained reason, the future is here described in the past tense.)

After drinking his breakfast, Dave began work. "Dave was an editor and feature writer for Trans Com News Service, one of the world's largest electronic news organizations. Although he routinely worked on stories of national and international events, he seldom left the apartment. His video screen gave him access to all of Trans Com's files. He could interview almost anyone in the world—from prime minister to Eskimo trader—via Vision-Phone."

By Vision-Phone, Dave interviewed "the minister of agriculture in Buenos Aires," composed his article, and then "activated the house monitor computer system," which reminded him "that Rent-A-Robot would be coming in to clean."

Ava, Dave's wife, worked in a factory. She did her work in a "control room" before "an enormous array of keyboards, video screens, and ranks and files of tiny lights." Her work was "kept track of" by a "central computer" known as "the front office." The members of "Ava's crew . . . were, of course, machines." "Although she was called a supervisor, she really did no supervision."

In the evening, the Nirays and their son, Billy, played electronic games on their video-screens.

This is a remarkable world in several respects. These people are apparently able to live an entire day without fulfilling directly any necessity of their lives. They do not take pleasure in any physical contact with anything or anybody. It is not recorded that they ever touch or speak to each other. Nor apparently, do they ever think a thought. Their entire mental life is devoted to acquiring things, getting promoted, and being electronically amused.

Although this world is enormously sophisticated technologically, it is biologically cruder and more irresponsible than our own. The Nirays drank "water recycled directly from sewage" because "there was really little choice. The wells had gone dry . . . years ago, and only an idiot would try to purify the stuff that came out of lakes and streams in the year 2001."

The engineers at Purdue assume that technology can be substituted for biology (as for everything else) with perfect adequacy and safety. There is not an ecological, economic, political, esthetic, or social consideration anywhere in the account. In this world, as we see from the Niray's job descriptions, words do not mean what they say: Ava is a supervisor, not because she supervises, but because she is *called* a supervisor. And knowledge has become simply news. No one needs to write or speak with authority. It is normal procedure for a reporter to write an article about a country he has never seen. Technology has thus replaced truth; it has perfected the public lie. (I am writing this, come to think of it, on Washington's *official* birthday. His *real* birthday is day after tomorrow.)

This "future" society is built exclusively on the twin principles of "convenience" and "control"—built, that is, on the dread of any kind of physical activity remotely classifiable as work. "Convenience," raised to this power, means the exchange of dependence (on oneself, on other people, on other creatures) for "control." It means not needing anything or anybody in particular. "Control" means, ultimately, *being* controlled, for in this world every room is a "control room," and no one is ever beyond control.

This "future" is so dismal, I think, because it is so nearly lifeless. The only living creatures, or the only ones on view, are humans, and humans are rigidly isolated from one another. They make no direct connections. They deal with each other, as they deal with the material world, only through technology. They live by "remote control." In nothing else I have read has the meaning of that phrase come so clear. Remote control is *pure* control—control without contact, without feeling, without fellow-feeling, therefore without satisfaction. Or it is without satisfaction to any but the totalitarian personality that enjoys control for its own sake.

And so the first question raised by the work of these fanciful engineers is: Where does satisfaction come from? They apparently think it comes from living in a state of absolute control and perfect convenience, in which one would never touch anything except push buttons.

The fact is, however, that a great many people have gladly turned off the road that leads to "Midland City, U.S.A." They are the home gardeners, the homesteaders, the city people who have returned to farming, the people of all kinds who have learned to do pleasing and necessary work with their hands, the people who have undertaken to raise their own children. They have willingly given up considerable amounts of convenience, and considerable amounts of control too, and have made their lives more risky and difficult than before.

Why? For satisfaction, I think. And where does satisfaction come from? I think it comes from contact with the materials and lives of this world, from the mutual dependence of creatures upon one another, from fellow feeling. But you cannot talk about satisfaction in abstract terms. There is no abstract satisfaction. Let me give an example.

Last summer we put up our second cutting of alfalfa on an extremely hot, humid afternoon. Our neighbors came in to help, and together we settled into what could pretty fairly be described as suffering. The hayfield lies in a narrow river bottom, a hill on one side and tall trees along the river on the other. There was no breeze at all. The hot, bright, moist air seemed to wrap around us and stick to us while we loaded the wagons.

It was worse in the barn, where the tin roof raised the temperature and held the air even closer and stiller. We worked more quietly than we usually do, not having breath for talk. It was miserable, no doubt about it. And there was not a push button anywhere in reach.

But we stayed there and did the work, were even glad to do it, and experienced no futurological fits. When we were done we told stories and laughed and talked a long time, sitting on a post pile in the shade of a big elm. It was a pleasing day.

Why was it pleasing? Nobody will ever figure that out by a "logical projection." The matter is too complex and too profound for logic. It was pleasing, for one thing, because we got done. That does not make logic, but it makes sense. For another thing, it was good hay, and we got it up in good shape. For another, we like each other and we work together because we want to.

And yet you cannot fully explain satisfaction in terms of just one day. Satisfaction rises out of the flow of time. When I was a boy I used to dread the hay harvest. It seemed an awful drudgery: the lifting was heavy and continuous; the weather was hot; the work was dusty; the chaff stuck to your skin and itched. And then one winter I stayed home and I fed out the hay we had put up the summer before. I learned the other half of the story then, and after that I never minded. The hay that goes up in the heat comes down into the mangers in the cold. That is when its meaning is clearest, and when the satisfaction is completed.

And so, six months after we shed all that sweat, there comes a bitter cold January evening when I go up to the horse barn to feed. It is nearly nightfall, and snowing hard. The north wind is driving the snow through the cracks in the barn wall. I bed the stalls, put corn in the troughs, climb into the loft and drop the rations of fragrant hay into the mangers. I go to the back door and open it; the horses come in and file along the driveway to their stalls, the snow piled white on their backs. The barn fills with the sounds of their eating. It is time to go home. I have my comfort ahead of me: talk, supper, fire in the stove, something to read. But I know too that all my animals are well fed and comfortable, and my comfort is enlarged in theirs. On such a night you do not feed out of necessity or duty. You never think of the money value of the animals. You feed and care for them out of fellow feeling, because you want to. And when I go out and shut the door, I am satisfied.

That leaves a lot unexplained. A lot is unexplainable. But the satisfaction is real. We can only have it from each other and from other creatures. It is not available from any machine. The "futurologists" do not see it in the future because they do not understand it now.

15

Home of the Free
(1978)

I was writing not long ago about a team of Purdue engineers who foresaw that by 2001 practically everything would be done by remote control. The question I asked—because such a "projection" *forces* one to ask it—was, *Where does satisfaction come from?* I concluded that there probably wouldn't be much satisfaction in such a world. There would be a lot of what passes for "efficiency," a lot of "production" and "consumption," but little satisfaction.

What I failed to acknowledge was that this "world of the future" is already established among us, and is growing. Two advertisements that I have lately received from correspondents make this clear, and raise the question about the sources of satisfaction more immediately and urgently than any abstract "projection" can do.

The first is the legend from a John Deere display at Waterloo Municipal Airport:

INTRODUCING SOUND-GARD BODY . . .
A DOWN TO EARTH SPACE CAPSULE.

New Sound-Gard body from John Deere, an "earth space capsule" to protect and encourage the American farmer at his job of being "Breadwinner to a world of families."

183

Outside: dust, noise, heat, storm, fumes.
Inside: all's quiet, comfortable, safe.
Features include a 4 post Roll Gard, space-age metals, plastics, and fibers to isolate driver from noise, vibration, and jolts. He dials 'inside weather', to his liking . . . he push buttons radio or stereo tape entertainment. He breathes filtered, conditioned air in his pressurized compartment. He has remote control over multi-ton and multi-hookups, with control tower visibility . . . from his scientifically padded seat.

The second is an ad for a condominium housing development:

HOME OF THE FREE.

We do the things you hate. You do the things you like. We mow the lawn, shovel the walks, paint and repair and do all exterior maintenance.

You cross-country ski, play tennis, hike, swim, work out, read or nap. Or advise our permanent maintenance staff as they do the things you hate.

Different as they may seem at first, these two ads make the same appeal, and they represent two aspects of the same problem: the widespread, and still spreading, assumption that we somehow have the right to be set free from anything whatsoever that we "hate" or don't want to do. According to this view, what we want to be set free from are the natural conditions of the world and the necessary work of human life; we do not want to experience temperatures that are the least bit too hot or too cold, or to work in the sun, or be exposed to wind or rain, or come in personal contact with anything describable as dirt, or provide for any of our own needs, or clean up after ourselves. Implicit in all this is the desire to be free of the "hassles" of mortality, to be "safe" from the life cycle. Such freedom and safety are always for sale. It is proposed that if we put all earthly obligations and the rites of passage into the charge of experts and machines, then life will become a permanent holiday.

What these people are really selling is insulation—cushions of technology, "space age" materials, and the menial work of other people—to keep fantasy in and reality out. The condominium ad says flat out that it is addressed to people who "hate" the handwork of household maintenance, and who will enjoy "advising" the people who do it for them; it is addressed, in other words, to those who think themselves too good to do work that other people are not too good to do. But it is a little surprising to realize that the John Deere ad is addressed to farmers who not only hate farming (that is, any physical contact with the ground or the weather or the crops), but also hate tractors, from the "dust," "fumes," "noise, vibration, and jolts" of which they wish to be protected by an "earth space capsule" and a "scientifically padded seat."

Of course, the only real way to get this sort of freedom and safety—to escape the hassles of earthly life—is to die. And what I think we see in these advertisements is an appeal to a desire to be dead that is evidently felt by many people. These ads are addressed to the perfect consumers—the self-consumers, who have found nothing of interest here on earth, nothing to do, and are impatient to be shed of earthly concerns. And so I am at a loss to explain the delay. Why hasn't some super salesman sold every one of these people a coffin—an "earth space capsule" in which they would experience no discomfort or inconvenience whatsoever, would have to do no work that they hate, would be spared all extremes of weather and all noises, fumes, vibrations, and jolts?

I wish it were possible for us to let these living dead bury themselves in the earth space capsules of their choice and think no more about them. The problem is that with their insatiable desire for comfort, convenience, remote control, and the rest of it, they cause an unconscionable amount of trouble for the rest of us, who would like a fair crack at living the rest of our lives within the terms and conditions of the real world. Speaking for myself, I acknowledge that the world, the weather, and the life cycle have caused me no end of trouble, and yet I look forward to putting in another forty or so years with them because they have also given me no end of pleasure and instruction. They interest me. I want to see them thrive on their own terms. I hate to see them abused and interfered with for the comfort and convenience of a lot of spoiled people who presume to "hate" the more necessary kinds of work and all the natural consequences of working outdoors.

When people begin to "hate" the life cycle and to try to live outside it and to escape its responsibilities, then the corpses begin to pile up and to get into the wrong places. One of the laws that the world imposes on us is that everything must be returned to its source to be used again. But one of the first principles of the haters is to violate this law in the name of convenience or efficiency. Because it is "inconvenient" to return bottles to the beverage manufacturers, "dead soldiers" pile up in the road ditches and in the waterways. Because it is "inconvenient" to be responsible for wastes, the rivers are polluted with everything from human excrement to various carcinogens and poisons. Because it is "efficient" (by what standard?) to mass-produce meat and milk in food "factories," the animal manures that once would have fertilized the fields have instead become wastes and pollutants. And so to be "free" of "inconvenience" and "inefficiency" we are paying a high price—which the haters among us are happy to charge to posterity.

And what a putrid (and profitable) use they have made of the idea of freedom! What a tragic evolution has taken place when the inheritors of the Bill of Rights are told, and when some of them believe, that "the home of the free" is where somebody else will do your work!

Let me set beside those advertisements a sentence that I consider a responsible statement about freedom: "To be free is precisely the same thing as to be pious, wise, just and temperate, careful of one's own, abstinent from what is another's, and thence, in fine, magnanimous and brave." That is John Milton. He is speaking out of the mainstream of our culture. Reading his sentence after those advertisements is coming home. His words have an atmosphere around them that a living human can breathe in.

How do you get free in Milton's sense of the word? I don't think you can do it in an earth space capsule or a space space capsule or a capsule of any kind. What Milton is saying is that you can do it only by living in this world as you find it, and by taking responsibility for the consequences of your life in it. And that means doing some chores that, highly objectionable in anybody's capsule, may not be at all unpleasant in the world.

Just a few days ago I finished up one of the heaviest of my spring jobs: hauling manure. On a feed lot I think this must be real drudgery even with modern labor-saving equipment—all that "waste" and no fields to put it on! But instead of a feed lot I have a small farm—what would probably be called a subsistence farm. My labor-saving equipment consists of a team of horses and a forty-year-old manure spreader. We forked the manure on by hand— forty-five loads. I made my back tired and my hands sore, but I got a considerable amount of pleasure out of it. Everywhere I spread that manure I knew it was needed. What would have been a nuisance in a feed lot was an opportunity and a benefit here. I enjoyed seeing it go out onto the ground. I was working some two-year-olds in the spreader for the first time, and I enjoyed that—mostly. And, since there were no noises, fumes, or vibrations the loading times were socially pleasant. I had some help from neighbors, from my son, and, toward the end, from my daughter who arrived home well rested from college. She helped me load, and then read *The Portrait of a Lady* while I drove up the hill to empty the spreader. I don't think many young women have read Henry James while forking manure. I enjoyed working with my daughter, and I enjoyed wondering what Henry James would have thought of her.

16

Going Back—
or Ahead—to Horses
(1980)

For several years now—ever since the beginning of the "energy crisis" and the concurrent rise in the price of draft horses and mules—some of our "agribusiness" experts have been proclaiming in what appears to be a kind of panic that it is "impossible to go back" to working horses on the farm. On their own terms, they are correct: we cannot change at once to a complete dependence on horses (nobody ever suggested that we could or should); and one man obviously cannot raise 300 acres of corn with a team. What the experts forget is that not everyone farms on their terms, and not everyone intends to.

While the experts have been raising such a fuss about the "impossibility" of farming with horses, a good many farmers scattered all over the country have quietly gone back to using them—just as thousands of Amish farmers have quietly *continued* to use them, with results pleasing to anybody who will look.

The farmers I know who have recently started using horses again are doing so because they can use them more cheaply for certain jobs than they can use a tractor. And for a lot of them the economic reason supports another reason just as important: they like horses, and like to work them.

The return to the use of horse or mule teams is easiest on diversified farms that are relatively small, and it is on such farms that they can most readily be seen as a solution to the high cost of fossil fuel energy. On such farms they are used for hauling, seed bed preparation, planting, mowing, spreading manure, etc. The farmer usually keeps his tractor to do the heavier jobs such as ground breaking and to use for belt or power takeoff work.

But work is also being found for teams on some of the larger farms, where energy costs are making the discrepancy between big tractors and little jobs increasingly hard to ignore: you don't need a hundred-horsepower machine to drag an end post or to haul out a load of hay or to empty a spreader load of manure. Many farmers are again discovering that a team is ideal for feeding stock in the winter. Unlike a tractor, a team will always start; it is not so quickly stopped by mud or snow; and it does far less damage to soft ground. Not least among the advantages is that, with a good team, feeding hay out on the ground is a one-person job—whereas with a tractor a hand is needed to drive and another to unload.

And so on many farms there is a place for a team or two—there always has been—and many of those places are again being filled.

I have been talking lately with Nick Coleman, who farms 182 acres near Sulphur, Kentucky, about this renewal of interest in working horses. Like a good many farmers who grew up using horses and who love them, Nick never did entirely quit using them; he always kept a horse or two around to do odd jobs and to cultivate his tobacco. But in the last two years his use of them has greatly increased. His goal is to do everything with his horses that can be done with them. Eventually, he says, he will have work for four horses; at present he has only a pair, a mare and a gelding. With these he does all his planting, cultivating, pasture clipping, manure hauling, feeding, fencing, and odd jobs.

Except for some raking and some hauling, he still harvests his hay with his tractors—a Massey-Ferguson 40 and a Case 570—which he also uses for power takeoff work. As soon as he acquires the necessary horses and can complete the adaptation of the machinery, he plans to reduce even further his dependence on the tractors.

When you ask him why, Nick will eventually talk about the cost of fuel—he figures that his increased dependence on horses has already cut his fuel consumption in half. But that is not his first reason. His first reason is more fundamental: "Pleasure! I like horses. I like to use them." He is right, I think, to insist on this, for it is necessarily the first rule: If you *don't* like them, don't use them.

He uses horses for the satisfaction of it, Nick says, but he does not leave it there; he considers that work horses belong inseparably to the kind of farming that is most satisfying. He uses them "for the satisfaction of farming right."

As a farmer, Nick is of the old-fashioned, conservative kind, unwilling to pay the costs and damages of extensive row cropping. He is mostly a grass farmer, and his produce mostly leaves the farm on the hoof. Of his 182 acres he has only eleven in row crops—six in corn, five in tobacco. He also raises eight acres of oats. That is, only about 10 percent of his land is broken in any year. The rest is kept in sod for hay and pasture for his fifty brood cows, safe both from erosion and from high production and maintenance costs. "And it's going to stay that way," Nick says.

As grazing animals, horses fit economically into such a program. They live off the land they work, and so, after purchase and occasional veterinary fees, cost no money. Nick figures that the expense of keeping a horse is about equal to that of keeping a cow. In addition to hay and oats, he thinks, his horses use about half an acre of pasture each—but that is improved pasture, well cared for, and grazed in rotation.

But there is still another reason that Nick prefers to farm with horses: "I like to work with machinery I can fix." As compared with tractor implements that are increasingly complicated, and difficult and expensive to repair, horse-drawn equipment is relatively simple in design, easy to understand, and repairable more often than not by the ordinary tools and know-how available on the farm. And not only is this equipment easier and cheaper to repair than tractor equipment, it needs repairing less often, because it is used at slower speeds.

I asked Nick what advice he had for anyone inexperienced who might want to use horses. "Find somebody who knows how," he said. "You can't learn it by yourself, and you can't read it out of a book. A book is all right, but you need to know something before you start reading."

Another farmer who does most of his work with horses is Maurice Telleen, whose sixty-acre place is on the outskirts of Waverly, Iowa, where he and his wife, Jeannine, edit *The Draft Horse Journal*. The main enterprise of the Telleen farm is a breeding flock of sixty to seventy-five Oxford ewes. In addition, Maury keeps a Guernsey milk cow, half a dozen or so saddle and draft horses, and usually four nursing calves. This year (1980) he is raising five acres of oats, six and a half acres of corn, about an acre and a half of tall sorghum, and ten acres of alfalfa. The rest of the farm is in pasture.

This, like Nick Coleman's, is a livestock farm, mostly in sod, and the work horses fit comfortably, almost naturally, into it. The present work force consists of three Percheron mares, and Maury considers three to be enough. That number pretty well assures him of having a pair to work should one of the mares be foaling or laid up for any other reason, and then he can use all three for such heavy jobs as breaking ground or disking.

Maury does not own a tractor, but like many farmers who work horses he will employ a tractor when it makes sense to do so. He routinely pays a tractor operator to bale his hay and do his combining, and will not hesitate to call in a tractor for other jobs as occasion or necessity demand. He is comforted to know that farming can be done without tractors, and he believes that more of it *will* be done without them, but he sees no sense in being a fanatic about it.

Maury figures that it costs about $150 to keep one of his draft mares for a year. Where he lives he has to expect the winter feeding to last 160 days. During that time his mares receive a maintenance ration that comes to a total of about a ton of hay per head (about eighty bales). In addition, they each receive a couple of ears of corn per day. In summer the horses are pastured and fed oats and a little hay as needed, according to the amount of work they do. Each horse is allotted about two acres of pasture.

The question of the costs and acreages required for the maintenance of draft horses is at present extremely difficult to answer. The answer will vary, sometimes greatly, from farm to farm, from farmer to farmer, and from horse to horse. But answers even from individual farmers are hard to come by, because draft horses are usually fed from the same cribs, hay mows, and pastures that feed other livestock—and no farmer that I know has ever had the inclination or the time to monitor exactly the feed given to any one horse. In my experience, however, the figures given by farmers who use horses have invariably been much lower than those given by the experts in their attempts to prove the "impossibility" of using horses.

For anyone wanting to start using horses on the farm, Maury had several suggestions:

1. Don't be carried away by the fad of bigness. Super-

sized horses will be more expensive than smaller ones both to buy and to feed.

2. Don't be carried away by youth and beauty. A horse that is sound, sensible, and well broke will serve a beginner far better—and be cheaper—than some beauty contestant that is "full of potential" or "has the makings of a good 'un."

3. If you are new to horses, buy broke geldings rather than mares. Geldings are more even-tempered than mares, and are usually cheaper. Beginners, especially if they have had no experience with horses of any kind should be cautious about the notion of a dual-purpose team—mares that will do your work while raising foals. This is one of the most attractive possibilities of farm horsepower. But the risks are greater with brood mares and the headaches more numerous. A brood mare may indeed do double duty—and she may die foaling.

4. Long-legged, "up-headed" draft horses are now the fashion, but do not necessarily make the best farm teams. A low-set, thick horse is usually an easy keeper.

5. When you are looking at horses to buy, check their mouths (for age, soundness of teeth, etc.) and their eyesight. And remember the horseman's saying, often repeated because it is true, that "a horse is no better than his feet."

6. Don't economize on harness, tongues, or trees. If any of these should break on a pull, that could cause a runaway, or worse.

7. In changing over to the use of horses, be guided by your own confidence and comfort. If you feel uneasy in starting a new job, or feel you don't know what you are doing, get help, or wait a while.

8. A logical first job for beginners is hauling manure. Another is hauling hay to livestock in winter. Neither job

calls for great precision in driving, and neither is as hurried or continuous as some other jobs. (Obviously, on any job, the steeper the ground the greater the risk.)

At present, with understanding, due caution, and some patience, you can get a usable team and a set of used tools for something in the neighborhood of $5,000. How much sense such an investment will make will depend, of course, on your income and needs. If you have a tight budget, a lot of work to do, and no time when you are not in a hurry, then a team of horses would probably just be a burden to you; you won't use them, and you will regret the cost and bother of keeping them. On the other hand, if you have a small farm, or if you have a lot of jobs that you know a team of horses can do as well or better or cheaper than a tractor, then this is an investment that merits serious attention.

But the sense of it depends too on your values and preferences. It is hard to make sense of something you don't enjoy. Farmers who have a use for horses and use them and enjoy using them are making sense. When going back makes sense, you are going ahead.

17

A Few Words
for Motherhood
(1980)

It is the season of motherhood again, and we are preoccupied with the pregnant and the unborn. When birth is imminent, especially with a ewe or a mare, we are at the barn the last thing before we go to bed, at least once in the middle of the night, and well before daylight in the morning. It is a sort of joke here that we have almost never had anything born in the middle of the night. And yet somebody must get up and go out anyway. With motherhood, you don't argue probabilities.

I set the alarm, but always wake up before it goes off. Some part of the mind is given to the barn, these times, and you can't put it to sleep. For a few minutes after I wake up, I lie there wondering where I will get the will and the energy to drag myself out of bed again. Anxiety takes care of that: maybe the ewe has started into labor, and is in trouble. But it isn't just anxiety. It is curiosity too, and the eagerness for new life that goes with motherhood. I want to see what nature and breeding and care and the passage of time have led to. If I open the barn door and hear a little bleat coming out of the darkness, I will be glad to be awake. My liking for that always returns with a force that surprises me.

These are bad times for motherhood—a kind of biological drudgery, some say, using up women who could do better things. Thoreau may have been the first to assert that people should not belong to farm animals, but the idea is now established doctrine with many farmers—and it has received amendments to the effect that people should not belong to children, or to each other. But we all have to belong to something, if only to the idea that we should not belong to anything. We all have to be used up by something. And though I will never be a mother, I am glad to be used up by motherhood and what it leads to, just as—most of the time—I gladly belong to my wife, my children, and several head of cattle, sheep, and horses. What better way to be used up? How else to be a farmer?

There are good arguments against female animals that need help in giving birth; I know what they are, and have gone over them many times. And yet—if the ordeal is not too painful or too long, and if it succeeds—I always wind up a little grateful to the ones that need help. Then I get to take part, get to go through the process another time, and I invariably come away from it feeling instructed and awed and pleased.

My wife and son and I find the heifer in a far corner of the field. In maybe two hours of labor she has managed to give birth to one small foot. We know how it has been with her. Time and again she has lain down and heaved at her burden, and got up and turned and smelled the ground. She is a heifer—how does she know that something is supposed to *be* there?

It takes some doing even for the three of us to get her into the barn. Her orders are to be alone, and she does all in her power to obey. But finally we shut the door behind her and get her into a stall. She isn't wild; once she is confined it isn't even necessary to tie her. I wash in a bucket of icy water and soap my right hand and forearm. She is quiet

now. And so are we humans—worried, and excited too, for
if there is a chance for failure here, there is also a chance for
success.

I loop a bale string onto the calf's exposed foot, knot the
string short around a stick which my son then holds. I press
my hand gently into the birth canal until I find the second
foot and then, a little further on, a nose. I loop a string
around the second foot, fasten on another stick for a hand-
hold. And then we pull. The heifer stands and pulls against
us for a few seconds, then gives up and goes down. We
brace ourselves the best we can into our work, pulling as
the heifer pushes. Finally the head comes, and then, more
easily, the rest.

We clear the calf's nose, help him to breathe, and then,
because the heifer has not yet stood up, we lay him on the
bedding in front of her. And what always seems to me the
miracle of it begins. She has never calved before. If she ever
saw another cow calve, she paid little attention. She has, as
we humans say, no education and no experience. And yet
she recognizes the calf as her own, and knows what to do
for it. Some heifers don't, but most do, as this one does.
Even before she gets up, she begins to lick it about the nose
and face with loud, vigorous swipes of her tongue. And all
the while she utters a kind of moan, meant to comfort, encourage, and reassure—or so I understand it.

How does she know so much? How did all this come
about? Instinct. Evolution. I know those words. I understand the logic of the survival of the fittest: good mothering
instincts have survived because bad mothers lost their
calves: the good traits triumphed, the bad perished. But
how come some are fit in the first place? What prepared in
the mind of the first cow or ewe or mare—or, for that matter, in the mind of the first human mother—this intricate,
careful, passionate welcome to the newborn? I don't know.
I don't think anybody does. I distrust any mortal who

claims to know. We call these animals dumb brutes, and so far as we can tell they are more or less dumb, and there are certainly times when those of us who live with them will seem to find evidence that they are plenty stupid. And yet, they are indisputably allied with intelligence more articulate and more refined than is to be found in any obstetrics textbook. What is one to make of it? Here is a dumb brute lying in dung and straw, licking her calf, and as always I am feeling honored to be associated with her.

The heifer has stood up now, and the calf is trying to stand, wobbling up onto its hind feet and knees, only to be knocked over by an exuberant caress of its mother's tongue. We have involved ourselves too much in this story by now to leave before the end, but we have our chores to finish too, and so to hasten things I lend a hand.

I help the calf onto his feet and maneuver him over to the heifer's flank. I am not supposed to be there, but her calf is, and so she accepts, or at least permits, my help. In these situations it sometimes seems to me that animals know that help is needed, and that they accept it with some kind of understanding. The thought moves me, but I am never sure, any more than I am sure what the cow means by the low moans she makes as the calf at last begins to nurse. To me, they sound like praise and encouragement—but how would I know?

Always when I hear that little smacking as the calf takes hold of the tit and swallows its first milk, I feel a pressure of laughter under my ribs. I am not sure what that means either. It certainly affirms more than the saved money value of the calf and the continued availability of beef. We all three feel it. We look at each other and grin with relief and satisfaction. Life is on its legs again, and we exult.

IV

18

A Rescued Farm
(1980)

At least since the early sixties, when strip mining became so extensive and destructive that it could no longer be ignored, there has been an almost continuous public effort to "regulate" it, and to "reclaim" the land afterwards. And there have been a lot of ideas, public and private, for using this reclaimed land. There has, I grant, been some success: there is a lot more regulation and reclamation now than there was fifteen years ago. But there is also a lot more strip-mined land than there was fifteen years ago; the government machinery necessary to regulate and reclaim is a lot bigger and more expensive than it was fifteen years ago; and no strip-mined land, however regulated and reclaimed, is as good as it was before—and, in human time, it is not going to be as good.

As for the ideas for using this land, I believe that nearly all of them have failed to materialize. My own opinion is that these ideas have been thought up too far from the strip mines. When the thinkers have got to the abandoned benches and spoil banks, the prospect has looked paralyzingly dismal, and they have looked for something easier to think about. At present, the principle user of exhausted strip mines is Nature, who is trying to grow weeds and bushes on them. Most of the human work done on strip mines—both to make and to unmake them—is still done by people who work for absentee owners and absentee governments, and who will not have to live with the consequences of their work.

I have often wondered what sort of human (as opposed to industrial or official) intention might finally turn toward these wasted places. And I had no answer until several months ago I received a letter from Wallace Aiken of East Palestine, Ohio: "The past several years I have been trying to reclaim some strip-mined land. I have a big bulldozer I use for this." The letter said "several years"—it wasn't written out of a pipe dream—and it contained information that suggested that this man knew what he was doing.

I replied by asking if I might come to see his work. The restoration of ruined land is a subject of great importance now in our country, where ruining land has been so lucrative and respectable a business, and I thought Wallace Aiken's methods and the extent of his success might therefore be of interest. But I also wondered what manner of man might, on his own, put "several years" into the reclamation of a strip mine. Mr. Aiken said that he would be pleased to show his farm and his work.

Wally Aiken is employed as a grinder in a foundry. He works the second shift so as to have the mornings free to work on his farm: "I'd rather be tired for their work than for mine." He lives with his mother on the outskirts of East Palestine in a good house built by his grandfather in 1912 for $1500. There is a family memory of a nearby hillside on which the townspeople pastured their milk cows, the children coming out night and morning to drive the cows home for milking. Now the hillside has been "developed," the hilly farmland beyond has been strip mined, and the children, having no cows to bring home, contract the sophistications of TV.

Wally's farm is half a mile from his house. Outside town you leave the blacktop and climb a steep haulroad corduroyed with railroad ties. At the top of the hill is a concrete block building surrounded by mowed lawn, a vegetable garden, big stacks of sawed firewood, and a machinery shed. From the hilltop you look down a long stretch of green hillside in a landscape grown over with trees and bushes. This opening is Wally Aiken's farm. Even from the height you can see that it is a restoration project, for the grass cover varies in density from one place to another, and at the far end of the opening Wally's enormous orange bulldozer sits in the midst of a patch of newly graded bare earth.

The block building was once the clubhouse of a local ham radio group. Wally was a member, and when the group split up he bought the clubhouse and the two acres of hilltop that went with it. That was his start in farming. Before that he had no agricultural experience.

For a while he raised hogs in pens now weed-grown on the slope below the clubhouse. But the next real step in his development as a farmer came when the adjoining forty-two acres came up for sale at a sheriff's auction. Wally bought it for $750—about $18 an acre. At the time he thought simply that it looked like a cheap way to own some land.

The land had been stripped for coal "on several different occasions in the past forty years." Wally, who is now thirty-two, has "no real idea what the place looked like before." What he had was a "farm" with a "surface like ripples on a pond"—the "ripples" being six to twelve feet high and covered with trees, bushes, briars, and vines. To walk across it would have been no casual undertaking. Nevertheless, it somehow got to be a farm in Wally's mind. All he had to do was correct the mayhem and negligence of his predecessors—a project, I think, that would have daunted a civil engineer, much less a would-be farmer who had never run a bulldozer.

He was a long way from farming. But he found the old bulldozer—an Allis-Chalmers HD 19—bought it for $2500, rebuilt the engine with the help of a more experienced friend, and started making his farm.

Though he has used the technology of the strip miners and some of their know-how, his methods have been opposite to theirs. They made rubble out of a farm. His aim has been to make a farm out of the rubble. To begin with, he wasted nothing. Before he started in with the bulldozer, he took out all the firewood and fence posts. The land was unquestionably a mess, but Wally's initial accomplishment seems to have been to respect it as it was; he would waste no part of the mess that he could see a way to use. And now that he has smoothed them out, he feels "a certain nostalgia toward those humps and undulations of earth." Clearing the land, he says, "gave me the first opportunity to come to know it." But he came to know it in the process of changing it. Now that most of the grading has been done, it no longer looks like the same place, and he has had to begin a new acquaintance with it.

The grading started in 1975, and that involved a sort of solitary apprenticeship as a bulldozer operator. The worst part of it, Wally says, was that from the driver's seat it is virtually impossible to see what you're doing: "I remember the first time I pushed dirt . . . I'd push ten feet or so, then hop down and look around front to see what I'd done." He still seems a little awed to think that so large a machine has to be run so much by guess—but that, he says, is the way you run it. "I'll always remember that old bulldozer roaring away and me sitting there trying to do something with it, and learning to guess."

It is hard to imagine how you would undo the damage of big machines except by a big machine, and so Wally has necessarily reconciled himself to the bulldozer. But he remains in a kind of conflict with it too. It is a powerful generalizer, and tends, just by its size and power, to work against his own governing impulse to take care of things, pay attention to details. It is too easy to be lazy when you are on the dozer: "It'll move anything. It's hard to save a log or a tree. I have to keep telling myself: 'Get off of this thing.'"

Wally has had a lot of dirt to move, a lot of uphill pushing, and the work has gone slowly. One plot of about an acre required 170 hours to backfill and grade. But now, in the sixth summer of the work, a small farm is, sure enough, appearing where the "ripples" of the old spoil banks have subsided beneath the dozer blade. Wally doesn't know exactly how many acres he has recovered, but it looked to be somewhere in the neighborhood of ten—which, added to the several acres of unstripped land on the property, will indeed make up a little farm, with plowland, pasture, and woodland, that will provide a subsistence and a modest surplus to sell.

Once the bulldozer work is finished on a given plot, Wally does what he calls the "groundwork"—picks up the larger stones, the tree branches, and the roots. And then he prepares a seedbed. This phase of the work needs to be carried out in haste: "The fewer trips the better." He sows rye along with a mixture of grasses and legumes: timothy, orchard grass, tall fescue, red clover, alsike, alfalfa, birdsfoot trefoil. The seed is put in with a cultipacker—which, he says, produces far better stands than he got before he bought it.

Once the seed is in the ground, Wally likes to cover it with a mulch of spoiled hay or manure. One year he hauled and spread eighty-five pickup loads of manure. When I visited Wally's place he showed me mulched and unmulched seedings in adjoining strips. The mulched strips were remarkably better.

But to give an idea of the way Wally thinks about his work and does it, I will quote his description of one of his problems:

> Since I was filling against a hillside, the finished grade was a slope. I knew I had to get some cover established as quickly as possible. I skipped picking up the stones, etc., and disked it lightly, then planted my grass-legume mixture. I then went over the whole thing with a cultipacker. I next spread out several hundred bales of spoiled and low-quality hay I bought. Well, it didn't rain, and it didn't rain, but the seedlings sprouted anyway just from the moisture in the ground. Weeks went by and still no rain. Then came hurricane Frederick. It rained and rained. I was afraid to look at my future pasture, but finally did. I was surprised to see that it had fared pretty well, considering the volume of rain we had had. It had washed in places, to be sure, but the damage was small. The washes were quickly filled with more hay. I think the hay mulch was the real lifesaver in this instance. The grass and legumes made several inches of growth before winter came.

My walk across Wally's remade farm began at a plot reclaimed the previous year. There was a good deal of bare ground showing through a stand of grasses and legumes that were obviously struggling for a roothold. And so I began in doubt. How would this pale mixture of subsoil and gravel ever support a sod? Who, after so much work, could be encouraged by this result? By the time we reached the oldest of the reclaimed plots, my doubts were gone. The ground was covered everywhere by a dense, thriving stand of pasture plants comparable to the best you would see anywhere. And underneath the sod was a brown, duffy layer of humus, where topsoil was building again. I was impressed to see that this layer was already thicker under a six-year-old sod than it was under the thirty- or forty-year-old thicket growth on the spoil banks.

And so the evidence is there. Wally Aiken is doing what he set out to do; he is making a farm out of the spoil left years ago by people who turned good land into money and smoke, in contempt of everything that might come after them. That they have had human successors at all here, where their destruction is bewildering and depressing, and in a nation where the care of such land is a mixture of bad habit and shoddy policy, is itself a kind of wonder. Wally Aiken has become their successor and their correction because, he says, "It is nice to have something to devote oneself to, to care about and be a part of." And his work has mended both the ruined land and (in himself at least) the human greed and folly that ruined it.

Why has he done it? I can only take him at his word—for what he has done is not "practical" or "economical," as things are now reckoned, and certainly not easy. He has done it out of devotion to a possibility once almost destroyed in his place, and now almost recovered. He has had a few things in his favor: the land is not too steep or too rocky, the soil is apparently rich in the necessary minerals, and he has had no trouble with acid drainage from the coal seams. But at the start that possibility lay as much in the mind and character of Wallace Aiken as in his land. The work, he says, has been satisfying all along, but "the final satisfaction will come when the place can produce, support animals, have fences, and begin to resemble some sort of agricultural operation. This is the goal I work toward."

19

An Excellent Homestead
(1979)

When you go to see Tom and Ginny Marsh you turn off a
busy road near Borden, Indiana, into an almost hidden entrance. You cross a stream, pass the end of a hedgerow of
autumn olives, drive past berry beds and a garden, and
suddenly you are struck by one of the most pleasing of real-
izations: you have come to a place that loving attention has
been paid to. Everywhere you look you see the signs of
care. You are in what appears to be a little cove, a wedge of
flat land tucked in against a wooded hill. The natural character of the place has been respected, and yet it has been
made to accommodate gracefully the various necessities of
a family's life and work.

There are gardens, berry patches, fruit trees, grape vines.
There are buildings, each one carefully designed and
placed: house, barn, sheds, a hog pen, a henhouse. And
there are a pottery workshop, display room, and kilns—for
both of the Marshes are potters and teachers of pottery
making at the University of Louisville. Their work is well
and widely known for its excellence, but one of their finest
accomplishments is the way they have joined their work
and their life, carrying on the most important aspects of
both here in the one place.

This connection is of the greatest importance to the Marshes: they have worked hard at it, and thought patiently about it. The relation between subsistence farming and pottery making, to begin with, is ancient and elemental. The Marshes grow most of their own food, and they make most of the dishes that they use to prepare and serve it. And they do each kind of work under the influence of the other: they farm with the exacting workmanship of artists; they make pottery with a good farmer's respect for earthly things.

Tom speaks of the "kinship" of farming and pottery making: "Both are arts; the products of both come from the earth and wind up on the dinner table; both are learned slowly. In both, you are working now for what will be in twenty years. You can't learn either one just by being *told*."

"If you are a good craftsman," he says, "you want to make the best possible dish or teapot—but you also want to build the best possible henhouse or clothesline pole." This understanding makes it possible to work wholeheartedly at whatever needs to be done. The Marsh homestead does not have the organization of a specialist enterprise. It is not partitioned into "work spaces" and "living spaces." "There is no disruption," Tom says, "in having to leave the studio to go do the mowing." And Ginny says: "A good cheese is as hard to make as a good pot."

When the Marshes started work here in 1970, the place was disused and neglected. The flat land where their buildings and garden now stand was just a sort of gravel bed with the creek wandering over it, and all grown up in briars. They straightened the creek and so salvaged the land they needed for building and food raising. They staked out a plot of the almost sterile soil and began to enrich it with heavy applications of sawdust and with green manure crops. They built their house and began to construct the other buildings they needed, working out the designs and doing the carpentry and masonry themselves. They set out fruit trees and berry bushes, and acquired a small laying flock. Along the new course of the creek they planted white and red pines, giving their place a wide margin set off by an autumn olive hedge. They dug a pond up on the hillside and stocked it with bass and bluegills. By December of 1975 they had cleared some hillside pasture, and so were ready for a milk cow. The next spring, because they had milk to spare, they started raising their own meat hogs.

By now they have an exemplary subsistence farm. Of their 12½ acres, all but four are wooded. Two of the cleared acres are in pasture. The rest of this space is taken up by buildings, garden, berry beds, and fruit trees. Because their acreage is so small, the Marshes have had to work carefully at the problems of design and scale. Everything had to be put in the right place or it would be in the way of something else. And it had to be the right size or they would run out of room.

Within these limits, they have thought competently, and worked well. They have considered the questions of scale, balance, proportion, unity, utility. They have tried to design and locate each of their buildings in the best relation to the land surface, the light, the weather, the other buildings, and the pattern of their own daily tasks. The result is that their homestead gives you something of the same pleasing sense of compact order that you get from looking at a Japanese garden: a small space has been made to contain a surprising number and variety of things, but there is no feeling of crowdedness because the proportions are right and everything is where it belongs.

One of the most pleasing qualities of this place is the unfailing appropriateness of its scale. Nothing is too big or too little. The henhouse, for instance, measures a little more than five feet by a little more than six. This accommodates, usually, ten hens and a rooster, a flock that provides more than enough eggs, and requires less trouble and expense, the Marshes say, than a dog. The hog pen is built on a concrete slab twenty feet long by eight feet wide. At one end there is an enclosure of four by eight feet, with a roof that lifts off for ease of cleaning. This is adequate for three hogs, though the Marshes prefer to keep only two. They have found it convenient to build their compost bins adjoining the hog pen. To load his stock, Tom has built a good, strong loading chute, the right size for a pickup truck and light enough to be carried by two people.

In all this, balance has been the aim or standard. Balance has defined the limits. The Marshes wanted a place large enough to provide a subsistence, but not too large to care for. They wanted to grow enough food so that they could keep what they needed and give some to friends. They wanted their life to have a margin of generosity, but not of waste.

More than by any other one thing, the Marshes think, their homestead was completed by their milk cow. She has, you might say, "tied the knot." She provides milk for their table, milk for a calf, and milk for the hogs. She makes manure, which the Marshes consider as valuable as her milk. She is utilizing land reclaimed from thicket; without her, it would be going to waste. And milking her morning and night, Tom says, gives their life a daily focus it did not have before. Ginny—characteristically, I think—wanted to take every possible advantage of having the cow, and she has become an accomplished maker of cheese and butter, and of pottery utensils for preparing and serving milk products.

The only disadvantage was that the Marshes occasionally needed to be away from home at milking time. That made them dependent on help from neighbors, but their younger neighbors did not know how to milk, and their

older neighbors, who knew how to milk, were out of practice or had arthritis in their hands. Tom solved the problem
with a forty-year-old milking machine that he restored and
set up in the barn. He is good at that sort of salvage, and it
has been important to the making of the place. For instance, his tractor, a Massey Ferguson 50, was made of
parts of two tractors, one that had been wrecked, another
that had burned.

The Marshes insist on the importance of understanding
that their homestead has been made slowly, not laid out according to a preconceived plan. The designing of their place
has been inseparable from their living in it. They have built
as their use of the place has defined possibilities and needs.

And this suggests another kinship between their life and
their work, farming and pottery, for their pottery too is
made to be used, not just looked at. Their favorite pots are
kitchen pots. They believe, as Tom says, that there is "a
profound beauty in function and utility." This gives their
life as well as their work, in Ginny's words, "a sense of
human kinship and communion with the earth."

This household and homestead belong where they are
because they have been fitted carefully and preservingly
into the life cycle. The human life of the place and the life
of its soil and its plants and animals revolve together. If the
Marshes see the completion of their place most clearly in
their milk cow, a visitor sees it most clearly in the house
when he sits down to a meal with Ginny and Tom and their
children. The house is cleanly proportioned, snug and comfortable, heated by a wood stove. There are books and a lot
of pottery. The dining table stands in the living room in
front of a large window which looks out at the garden. The
living room opens directly into the kitchen. As in most
homestead households, food here is nobody's specialty. It is
the work of both husband and wife—another way of tying

the knot. The table is spread with excellent food, beautifully served on beautiful dishes. Eating, you feel the cycle
turn, completing itself yet again. The cow eats, the hogs
eat, the chickens eat, the people eat. The life of the place
comes in as food, returns as fertility, comes in as energy,
returns as care.

20

Elmer Lapp's Place
(1979)

The thirty cows come up from the pasture and go one by one into the barn. Most of them are Guernseys, but there are also a few red Holsteins and a couple of Jerseys. They go to their places and wait while their neck chains are fastened. And then Elmer Lapp, his oldest son, and his youngest daughter go about the work of feeding, washing, and milking.

In the low, square room, lighted by a row of big windows, a radio is quietly playing music. Several white cats sit around waiting for milk to be poured out for them from the test cup. Two collie dogs rest by the wall, out of the way. Several buff Cochin bantams are busily foraging for whatever waste grain can be found in the bedding and in the gutters. Overhead, fastened to the ceiling joists, are many barn swallow nests, their mud cups empty now at the end of October. Two rusty-barrelled .22 rifles are propped in window frames, kept handy to shoot English sparrows, and there are no sparrows to be seen. Outside the door a bred heifer and a rather timeworn pet jenny are eating their suppers out of feed boxes. Beyond, on the stream that runs through the pasture, wild ducks are swimming. The shadows have grown long under the low-slanting amber light.

This is a farm of eighty-three acres that has been in the Lapp family since 1915, five years before Elmer Lapp was born, and he has been here all his life. Three years ago a new house was built for Mr. Lapp's oldest son, who is his farming partner, father and son doing all the carpentry themselves. Except for the four or five days a month that the son works off the farm, the two households take their living from this place, plus fourteen acres of rented pasture and forty acres of hay harvested on the shares on a farm some distance away. They are farming then, all told, 117 acres.

Because this farm is in Lancaster County, Pennsylvania, in an enclave of Amish and Mennonite farms that has become a "tourist attraction," the Lapps are able to supplement their agricultural income by selling farm tours, chicken barbecue, and homemade ice cream to busloads of school children and tourists. But as profitable a sideline as this undoubtedly is, it should not distract from the economic and ecological good health of the farm operation itself. At a time when so many small farms are struggling or failing, it may be easy to suspect that this farm survives by dependence on the tourist industry. I do not think so. Here, at least, the opposite would seem to be true: the sideline succeeds because the main enterprise is a success.

Standing in the stanchion barn while the cows are being milked, I am impressed by how quietly the work is done. No voice is raised. There is never a sudden or violent motion. Although the work is quickly done, no one rushes. And finally comes the realization that the room is quiet because it is orderly: all the creatures there, people and animals alike, are at rest within a pattern deeply familiar to them all. That evening and the day following, as I extend my acquaintance with the farm and with Elmer Lapp's understanding of it, I see that quiet chore time as a nucleus or gathering point in a pattern that includes the whole farm.

The farm is thriving because what I would call its structural problems have been satisfactorily solved. The patterns necessary to its life have been perceived and worked out.

The Commercial Pattern

In its commercial aspect, this is a livestock farm. Its crops are not grown to sell, but to feed animals. The main enterprises are the thirty-cow dairy, and eleven Belgian brood mares.

Mr. Lapp's dairy herd is made up mainly of Guernseys because, he says, "Big cows eat too much." And the richer milk of the Guernseys brings a premium price. His few Holsteins are red ones, because their milk is richer than that of the blacks. Their milk "tests with the Guernseys'," Mr. Lapp says.

He now sells manufacturing milk to the people who make Hershey chocolate. He used to ship Grade A, but quit, he says, because "The Grade A guys got under my hide. You could never satisfy them. They always wanted something else." At several points in our conversation Mr. Lapp showed this sort of independence. He is not a man to put up long with anything he does not like. And this, again, I take as an indication of his success as a farmer. He is independent because he can afford to be.

At present, in addition to the thirty milking cows, he has twelve heifers, six of which he has just started on the bucket. He likes to have a couple of heifers coming fresh each year. He sells his bull calves as babies. His heifer calves are started on milk replacer which he considers better for the purpose than milk. They are given two quarts at a feeding.

When I ask Mr. Lapp what a farmer could expect to make from a farm of this size, managed as this one is, he replies by saying that he sells $20,000 to $30,000 worth of milk each year. Last year his dairy grossed $25,000.

I ask him how much of that was net.

He can't tell me exactly, he says. He bought $5,000 worth of supplements, but that included extra feed for his chickens, horses, and calves. And, of course, some of the expense was offset by the sale of bull calves and heifers. Aside from this information, he describes his income by saying "I pay taxes."

Mr. Lapp offers no information about his income from his horses. But the market for draft horses is booming, and one must suppose that the Lapp farm is sharing in the payoff. Last year Mr. Lapp sold nine head. This past season he has bred eleven mares. He also has an income from his stallion who serves, he says, "all the outside mares I can handle." Besides the brood mares and the stallion, he presently has on hand a two-year-old filly, two yearling fillies, two yearling stud colts, and two foals.

He prefers the draftier type of Belgians, but wants them long legged enough to walk fast, and because he works his horses he is attentive to the need for good feet. Along with those practical virtues, he likes his horses to show a good deal of refinement, and in selecting breeding stock pays particular attention to heads and necks. Among his mares are several that are half or full sisters, and this gives his horses a very noticeable uniformity of both color and conformation.

Because for some reason his land will not produce oats of satisfactory quality, Mr. Lapp grows barley for his horses. If barley was good enough horse feed for King Solomon, he says, it is good enough for him. He crimps or grinds the barley and adds molasses.

Unlike many horsemen, Mr. Lapp has no elaborate lore or procedure for breeding mares. He serves a mare only once, on whatever day he notices that she is in heat. And he sees no sense in pregnancy tests or examinations. Even so, he says, he has no trouble getting mares to conceive—or cows either, except with artificial insemination.

But just because his major income is from dairy cows

and brood mares, Mr. Lapp does not shut his eyes to other opportunities. "You stay awake," he says. He knows what will sell, and so far as his place and time allow he has it for sale. He feeds three hundred guineas at a time in a small loft. He raises and sells collie pups. He sells his surplus of eggs and honey. Even the barn cats contribute their share of income, for when he gets too many he sells the surplus at the local sale barn.

The Pattern of Subsistence

Though the Lapp farm is commercially profitable its balance sheet would fall far short of accounting for the life of the place, or even for its economy.

Elmer Lapp is eminently a traditional farmer in the sense that his farm is his home, his life, and his way of life—not just his "work place" or his "job." For that reason, though his farm produces a cash income, that is not all it produces, and some of what it produces cannot be valued in cash.

In obedience to traditional principle, the Lapps take their subsistence from the farm, and they are as attentive to the production of what they eat as to the production of what they sell. The farm is expected to make a profit, but it must make sense too, and a part of that sense is that it must feed the farmers. And so a pattern of subsistence joins, and at certain points overlaps, the commercial pattern.

For instance, the Lapps drink their own milk. I know that a lot of dairying families buy their milk at the grocery store, and so I ask Mr. Lapp why he doesn't buy milk for his own household.

He answers unhesitatingly: "I don't like that slop."

He also grows a garden. He has an orchard of apple, peach, and plum trees for fruit, and for blossoms for his bees. He is feeding four hogs, bought cheaply because they were runts, to slaughter for home use. He slaughters his own beef, and produces his own poultry, eggs, and honey.

He is also aware that the pattern of subsistence is a community pattern. He says, for instance, that he deals with the little country stores rather than the supermarkets in the city. The little country stores support the life of the community, whereas the supermarkets support "the economy" at the expense of communities.

The Patterns of Soil Husbandry

Underlying the patterns of the farm's productivity is a stewardship of the soil at all points knowledgeable, disciplined, and responsible. And this stewardship, necessarily, has evolved its own appropriate patterns.

In any year, Mr. Lapp will have twenty-two acres in corn (twelve for silage, ten to husk), twenty-five acres in clover or alfalfa, ten acres in barley or rye, and the rest in permanent pasture. The rotation is, mainly, as follows:

First year: Corn for husking.
Second year: Silage corn.
Third year: Barley, planted in preceding fall, with clover and timothy sowed broadcast onto frozen ground in spring. After the barley is harvested, the field produces one cutting of hay.
Fourth year: Clover and timothy (two cuttings).
Fifth year: Back to corn.

This pattern is varied in two ways. Where alfalfa is sowed instead of clover, the field is left in sod for three or four years instead of two. And when rye is sowed instead of barley, the rye is flail-chopped in the bloom and baled for bedding, and the land is returned to silage corn the same year.

The whole farm is covered with manure each year, at a rate, Mr. Lapp figures, of about eight tons per acre. And care is taken to get the manure on at the right time. I ask if this use of manure did not reduce the need for commercial fertilizer. "I don't buy any fertilizer," Mr. Lapp says. (He does use an herbicide on his corn fields, but only because the time when corn needs cultivation is also the time when he is busiest with tours.)

The present system of rotation and fertilization has been in use on this farm, Mr. Lapp says, "as long as I remember." But he himself, with the county agent's help, laid the farm off in three-acre strips to help control runoff and erosion. Yet even though soil conservation can to a considerable extent be formalized in set patterns of layout and rotation, there is still a need for vigilance and intelligent improvisation. This fall, for instance, the barley is coming on too late to provide good winter protection. As a remedy, Mr. Lapp says, he will cover the barley fields with strawy manure on the first morning the ground is frozen. That will protect the fields through the winter without smothering the barley.

One of the best ways to measure the quality of soil husbandry and the richness of soil on a farm is to look at its first-year hayfields. How quickly will clover and grass make a sod after the land has been row cropped? How healthy and productive is it? The height, density, color, and uniformity of the plants all have a tale to tell. Mr. Lapp leads the way up past his garden to a four-acre hayfield that is good in all respects. It was sowed in the spring to red clover, timothy, and a little alsike for the bees. The barley was taken off in July. And then in early October the field was mowed for hay, yielding 400 bales. Next year, it may reasonably be expected to yield 800–1000 bales on the first cutting, and 500–600 on the second.

Two Kinds of Horsepower
When Elmer Lapp was still just a boy, his father, recognizing a gift in him, gave him the colts to work.

"Made you a little proud?" I say.

He grins and nods. "I guess it did a little bit."

Because he is a capable horseman and likes horses, he has never quit using them—although he has certain uses for a tractor as well. "I'd rather drive horses than a tractor," he says. "I have them here, they're eating, so I might as well use them. I'm doing my work while I'm having pleasure. If I didn't enjoy it I wouldn't do it."

He uses a tractor for what a tractor does best, and horses for what they do best, keeping in mind always the scale of his operation. "On a small farm," he says, "you don't need expensive equipment." And he seems immune to the horsepower intoxication that leads so many small farmers to buy larger tractors than they need. He paid $2000 for a John Deere 60 twenty years ago, and is still using it. It will pull a three-bottom plow. When he needs a tractor for an occasional heavier job, such as silo filling, he hires a larger one. He does all his plowing and hay baling with his tractor, and uses it to load manure. He uses his horses to spread manure, plant corn, clip pasture, rake and haul hay. If he is "not pushed too hard," he uses them also in seedbed preparation. He is sure that he gets this work done cheaper with horses than with a tractor—even setting aside the value of their colts.

He says that rubber-tired equipment is far easier on horses than the steel-tired, because the tires absorb much of the shock when working over rough ground. And he dislikes wide hitches largely because they too are hard on horses. On an eight-foot tandem disk he will hitch two in front and three behind—or, if the footing is solid and the going relatively easy, he will work as many as four abreast. He says that he sees far too much mistreatment of horses through ignorance and indifference—something he resents and tries, so far as he can, to correct.

The use of the horses, whose feed is grown on the farm, greatly extends Mr. Lapp's dependence on solar energy, and greatly reduces his dependence on increasingly expensive fossil fuel energy. The tractor is used to supplement the energy already available on the farm.

In addition to the two varieties of horsepower, the farm makes a small use of waterpower. The stream is dammed and the impounded water used to turn a small water wheel which, in turn, works a water pump. It is another manifestation of this farm's thriftiness. Mr. Lapp looks at the escaping water with some regret: "That's all going to waste."

A Well-Planned Barn
The Lapps are just completing a small barn that is a good example of the care and the sense of order that have gone into the making of their farm.

This is a "bank barn" with a drive-in loft, approximately thirty by forty-eight feet. The lower story is a feeding area that will accommodate five hundred guineas in the summer and twelve heifers and perhaps as many young horses in the winter. It is divided across the middle by a feed bunk which extends out into a lot.

The upper story will have a corn crib across each end, eight feet wide by fourteen deep. The area in the center will be for storage of hay and equipment. The cribs are to be ventilated by lattices along the lower part of the outside walls. Outside, these lattices will be sheltered by awnings, four feet wide on one end, but on the other end ten feet wide to provide yet more shelter for equipment.

All possibilities of site, shape, and use have been considered.

The Ecological Pattern
Concerned as he is that the usable be put to use, that there be no waste, still there is nothing utilitarian or mechanistic about Mr. Lapp's farm—or his mind. His aim, it seems, is

not that the place should be put to the fullest use, but that it should have the most abundant life. The best farmers, Sir Albert Howard said, imitate nature, not least in the love of variety. Elmer Lapp answers to that definition as fully as any farmer I have encountered. Like nature herself, he and his family seem preoccupied with the filling of niches.

Driving into the place, one is aware before anything else that wherever flowers can be grown flowers are growing; beds and borders are everywhere. The barn swallow nests in the milking barn are not there just by happenstance; little wooden steps have been nailed to the joists to encourage them to nest there. Elmer Lapp has defended them against milk inspectors—"If those barn swallows go, I'm going somewhere else with my milk"—and against the cats, which he pens up during the nesting season, "if they get nasty."

Among the wild creatures, he seems especially partial to birds. Wild waterfowl make themselves peacefully at home along his pasture stream, and he speaks of his failure to attract martins with obvious grief. One can justify the existence of birds by "insect control," but one can also like them. Elmer Lapp likes them. His one acknowledged regret about his place is that it doesn't have a wood lot. He could use the firewood; he would also like the wild creatures it would attract. Above his row of beehives is a border of sudan grass that he has let go to seed for the birds.

He likes too the buff Cochin bantams that live in the milking barn and the stable—they scatter the manure piles and so keep flies from hatching—and the goldfish who live in the drinking trough and keep the water clean. Walking around the place, I keep being surprised by some other creature that has found room and board there, and is contributing a little something—maybe only pleasure—in return: peafowl, wild turkeys, pigeons, a pair of bobwhites.

For a man giftedly practical, Mr. Lapp justifies what he has and does remarkably often by his *likes*. One finally realizes that on the Lapp farm one is surrounded by an abounding variety of lives that are there, and are thriving there, because Elmer Lapp *likes* them. And from that it is only a step to the realization that the commercial enterprises of the farm are likewise there, and thriving, because he likes them too. The Belgians and the Guernseys are profitable, in large part, because they were liked *before* they were profitable. Mr. Lapp is as fine a farmer as he is because liking has joined his intelligence intricately to his place.

And that is why the place makes sense. All the patterns of the farm are finally gathered into an ecological pattern; it is one "household," its various parts joined to each other and the whole joined to nature, to the world, by liking, by delighted and affectionate understanding. The ecological pattern is a pattern of pleasure.

21

A Talent
for Necessity
(1980)

In the days when the Southdown ram was king of the sheep pastures and the show-ring, Henry Besuden of Vinewood Farm in Clark County, Kentucky, was perhaps the premier breeder and showman of Southdown sheep in the United States. The list of his winnings at major shows would be too long to put down here, but the character of his achievement can be indicated by his success in showing carload lots of fat lambs in the Chicago International Livestock Exposition. Starting in 1946, he sent eighteen carloads to the International, and won the competition twelve times. "I had 'em fat," he says, remembering. "I had 'em good." Such was the esteem and demand for his stock among fellow breeders that in 1954 he sold a yearling ram for $1200, then a record price for a Southdown.

One would imagine that such accomplishments must have rested on the very best of Bluegrass farmland. But the truth, nearly opposite to that, is much more interesting. "If I'd inherited good land," Henry Besuden says, "I'd probably have been just another Bluegrass farmer."

What he inherited, in fact, was 632 acres of rolling land, fairly steep in places, thin soiled even originally, and by the time he got it, worn out, "corned to death." His grandfather would rent the land out to corn, two hundred acres at a time, and not even get up to see where it would be planted—even though "it was understood to be the rule that renters ruined land." By the time Henry Besuden was eight years old both his mother and father were dead, and the land was farmed by tenants under the trusteeship of a Cincinnati bank. When the farm came to him in 1927, it was heavily encumbered by debt and covered with gullies, some of which were deep enough to hide a standing man.

And so Mr. Besuden began his life as a farmer with the odds against him. But his predicament became his education and, finally, his triumph. "I was lucky," he told Grant Cannon of *The Farm Quarterly* in 1951. "I found that I had some talent for doing the things I *had* to do. I *had* to improve the farm or starve to death; and I *had* to go into the sheep business because sheep were the only animals that could have lived off the farm."

Now seventy-six years old and not in the best of health, Mr. Besuden has not owned a sheep for several years, but he speaks of them with exact remembrance and exacting intelligence; he is one of the best talkers I have had the luck to listen to. How did he get started with sheep? "I was told they'd eat weeds and briars," he says, looking sideways through pipesmoke to see if I get the connection, for the connection between sheep and land is the critical one for him. The history of his sheep and the history of his farm are one history, and it is his own.

Having only talent and necessity—and unusual energy and determination—Mr. Besuden set about the restoration of his ravaged fields. There was no Soil Conservation Service then, but a young man in his predicament was bound to get plenty of advice. To check erosion he first tried building rock dams across the gullies. That wasn't satisfactory; the dams did catch some dirt, but then the fields were marred by half buried rock walls that interfered with work. He tried huge windrows of weeds and brush to the same purpose, but that was not satisfactory either.

Some of the worst gullies he eventually had to fill with a bulldozer. But his main erosion-stopping tool turned out, strangely enough, to be the plow; the tool that in the wrong hands had nearly ruined the farm, in the right hands healed it. Starting at the edge of a gulley he would run a backfurrow up one side and down the other, continuing to plow until he had completed a sizable land. And then he would start at the gulley again, turning the furrows inward as before. He repeated this process until what had been a ditch had become a saucer, so that the runoff, rather than concentrating its force in an abrasive torrent, would be shallowly dispersed over as wide an area as possible. This, as he knew, had been the method of the renters to prepare the gullied land for yet another crop of corn. For them, it had been a temporary remedy; he made it a permanent one.

Nowadays Kentucky fescue 31 would be the grass to sow on such places, but fescue was not available then. Mr. Besuden used small grains, timothy, sweet clover, Korean lespedeza. He used mulches, and he did not overlook the usefulness of what he knew for certain would grow on his land—weeds: "Briars are a good thing for a little hollow." In places he planted thickets of black locust—a native leguminous tree that would serve four purposes: hold the land, encourage grass to grow, provide shade for livestock, and produce posts. But his highest praise is given to the sweet clover which he calls "the best land builder I've ever run into. It'll open up clay, and throw a lot of nitrogen into the ground." The grass would come then, and the real healing would start.

Once the land was in grass, his policy generally was to leave it in grass. Only the best-laying, least vulnerable land was broken for tobacco, the region's major money crop then as now. Even today, I noticed, he sees that his fields are plowed very conservatively. The plowlands are small and carefully placed, leaving out thin places and waterways.

The basic work of restoration continued for twenty-three years. By 1950 the scars were grassed over, and the land was supporting one of the great Southdown flocks of the time. But it was not healed. What was there is gone, and Henry Besuden knows that it will be a long time building back. "'Tain't in good shape, yet," he told an interviewer in 1978.

And so if Mr. Besuden built a reputation as one of the best of livestock showmen, the focus of his interest was nevertheless not the show-ring but the farm, It would be true, it seems, to say that he became a master sheepman and shepherd as one of the ways of becoming a master farmer. For this reason, his standards of quality were never frivolous or freakish, as show-ring standards have sometimes been accused of being, but insistently practical. He never forgot that the purpose of a sheep is to produce a living for the farmer and to put good meat on the table: "When they asked me, 'What do you consider a perfect lamb?' I said, 'One a farmer can make money on!' The foundation has to be the commercial flock." And he wrote in praise of the Southdown ram that "he paid his rent."

But it was perhaps even more characteristic of him to write in 1945 that "one very important thing is that sheep are land builders," and to plead for their continued inclusion in farm livestock programs. He had seen the handwriting on the wall: the new emphasis on row cropping and "production" which in the years after World War II would radically alter the balance of crops and animals on farms, and which, as he feared, would help to destroy the sheep business in his own state. (In 1947, Mr. Besuden's county of Clark had twenty-four breeding flocks of Southdowns, and 30,000 head of grade ewes. That is more than remain now in the whole state of Kentucky.) What he called for instead—and events are rapidly proving him right—was "a long-time program of land building" by which he meant a way of farming based on grass and forage crops, which would build up and maintain reserves of fertility. And in that kind of farming, he was prepared to insist, because he knew, sheep would have an important place.

"I think," he wrote in his series of columns, "Sheep Sense," published in *The Sheepman* in 1945 and 1946, "the fertilizing effect of sheep on the farm has never received the attention it deserves. As one who has had to farm poor land where the least amount of fertilizer shows up plainly, I have noticed that on land often thought too poor for cattle the sheep do well and in time benefit the crops and grass to such an extent that other stock can then be carried. I have seldom seen sheep bed down for the night on anything but high land, and their droppings are evenly scattered on the pasture while grazing, so that no vegetation is killed."

What he wanted was "a way of farming compatible with nature"; this was the constant theme of his work, and he followed it faithfully, both in his pleasure in the lives and events of nature, and in his practical solutions to the problems of farming and soil husbandry. He was never too busy to appreciate, and to praise, the spiritual by-products, as he called them, of farm life. Nor was he too busy to attend to the smallest needs of his land. At one time, for example, he built "two small houses on skids," each of which would hold twenty-five bales of hay. These could be pulled to places where the soil was thin, where the hay would be fed out, and then moved on to other such places. (In the spring they could be used to raise chickens.)

"It's good to have Nature working for you," he says. "She works for a minimum wage." But in reading his "Sheep Sense" columns, one realizes that he not only did not separate the spiritual from the practical, but insisted that they cannot be separated: "This thing of soil conservation involves more than laying out a few terraces and diversion ditches and sowing to grass and legumes, it also involves the heart of the man managing the land. If he loves his soil he will save it." Once, he says, he thought of numbering his fields, but decided against it—"That didn't seem fair to them"—for each has its own character and potential.

As a rule, he would have 400 head of ewes in two flocks—a flock of registered Southdowns and a flock of "Western" commercial ewes. After lambing, he would be running something in the neighborhood of 1000 head. To handle so many sheep on a diversified farm required a great deal of care, and Mr. Besuden's system of management, worked out with thorough understanding and attention to detail, is worth the interest and reflection of any raiser of livestock.

It was a system intended, first of all, to get the maximum use of forage. This rested on what he understands to be a sound principle of livestock farming and soil conservation, but it was forced upon him by the poor quality of his land. He had to keep row cropping to a minimum, and if that meant buying grain, then he would buy it. But he did not buy much. He usually fed, he told me, one pound of corn per ewe per day for sixty days. But in "Sheep Sense" for December 1945, he wrote: "One-half pound grain with three pounds legume hay should do the job, starting with the hay and adding the grain later." He creep-fed his early lambs, but took them off grain as soon as pasture was available. In "Sheep Sense," March 1946, he stated flatly that "creep-feeding after good grass arrives does not pay."

Grain, then, he considered not a diet, but a supplement, almost an emergency ration, to assure health and growth in the flock during the time when he had no pasture. It must be remembered that he was talking about a kind of sheep bred to make efficient use of pasture and hay, and that the market then favored that kind. In the decades following World War II, cheap energy and cheap grain allowed interest to shift to the larger breeds of sheep and larger slaughter lambs that must be grain fed. But now with the cost of energy rising, pushing up the cost of grain, and the human consumption of grain rising with the increase of population, Henry Besuden's sentence of a generation ago resounds with good sense: "Due to the shortage of grain throughout the world, the sheep farmer needs to study the possibilities of grass fattening."

Those, anyhow, were the possibilities that *he* was studying. And the management of pasture, the management of sheep *on* pasture, was his art.

In the fall he would select certain pastures close to the barn to be used for late grazing. This is what is now called "stockpiling"—which, he points out, is only a new word for old common sense. It was sometimes possible, in favorable years, to keep the ewes on grass all through December, feeding "very little hay" and "a small amount of grain." Sometimes he sowed rye early to provide late fall pasture and so extend the grazing season.

His ewes were bred to lamb in January and February. He fed good clover or alfalfa hay, and from about the middle of January to about the middle of March he gave the ewes their sixty daily rations of grain. In mid March the grain feeding ended, and ewes and lambs went out on early pasture of rye which had been sown as a cover crop on the last year's tobacco patches. "A sack of Balboa rye sown in the early fall," he wrote, "is worth several sacks of feed fed in the spring and is much cheaper." From the rye they went to the clover fields where tobacco had grown two years before. From the clover they were moved onto the grass pastures. The market lambs were sold straight off the pastures, at eighty to eighty-five pounds, starting the first of May.

After fescue became available, Mr. Besuden made extensive use of it in his pastures. But he feels that this grass, though an excellent land conserver, is not nutritious or palatable enough to make the best sheep pasture, and so he took pains to diversify his fescue stands with timothy and legumes. His favorite pasture legume is Korean lespedeza, though he joins in the fairly common complaint that it is less vigorous and productive now than it used to be. He has also used red clover, alsike, ladino, and birdsfoot trefoil. He says that he had trouble getting his ewes with lamb in the first heat when they were bred on clover pastures, but that he never had this trouble on lespedeza.

His pastures were regularly reseeded to legumes, usually in March, the sheep tramping in the seed, and he found this method of "renovation" to be as good as any. The pastures were clipped twice during the growing season, sometimes oftener, to keep the growth vigorous and uniform.

The key to efficient management of sheep on pasture is paying attention, and it was important to Mr. Besuden that he should be on horseback among his sheep in the early mornings. The sheep would be out of the shade then, grazing, and he could study their condition and the condition of the field. He speaks of the "bloom" of a pasture, referring to a certain freshness of appearance made by new, tender growth sprigging up through the old. When that bloom is gone, he thinks, the sheep should be moved. The move from a stale pasture to a fresh one can lengthen the grazing time by as much as two hours a day. He believes also that lambs do best when the flock is not too large. That is because sheep tend to bunch together when grazing, the least vigorous lambs coming last and having to feed on grass mouthed over and rejected by the others. He saw to it that his pastures were amply provided with shade, and he knew that the shade needed to be well placed: "I think the best lamb-growing pastures I have are the ones where the shade is close to the water. I have seen times during July and August when sheep would not leave the shade and go to water if the shade and water happened to be at opposite ends of a large field."

The crisis of the shepherd's year, of course, is lambing time. That is the time that the year's work stands or falls by. And because it usually takes place in cold weather, the success of lambing is almost as dependent on the shepherd's facilities as on his knowledge. The lambing barn at Vinewood is an instructive embodiment of Mr. Besuden's understanding of his work and his gift for order. He gives a good description of it himself in one of his columns:

"Practically all the lambing here at Vinewood in recent years has been in a barn especially made for the purpose, shiplap (tongue groove) boxing with a low loft and a window in each bent. The east end of the barn [away from the prevailing winds] is rarely ever closed, a gate being used. Often in extremely cold weather the temperature can be raised fifteen or twenty degrees by the heat from the sheep. Some thirty feet out in the front and extending the width of the barn [is] a heavy layer of rock. . . . This prevents the muddy place that often appears at the barn door and . . . pulls at the sheep as they walk through it, causing slipped lambs. Also at the entrance . . . a locust post is half embedded across the door. This serves as a protection in case of dogs trying to dig under the door or gate and helps to hold the bedding in the barn as the sheep go out. Any kind of a sill that is too high or causes the heavy-in-lamb ewes to jump or strain to cross is too risky."

The barn is admirably laid out, with pens, chutes, and gates to permit the feeding, handling, sorting, and loading of a large number of sheep with the least trouble. There were lambing pens for forty ewes. There was also a small room with pens that could be heated by a stove. Above each pen was a red wooden "button" that could be turned down to indicate that a ewe was near to lambing or for any other reason in need of close attention. These were used when Mr. Besuden had an experienced helper to share the nighttime duty with him. "They saved a lot of cold midnight talk," he says.

But experienced help was not always available, and then he would have to work through the days and nights of lambing alone. Staying awake would get to be a problem. Sometimes, sitting beside one of the pens, waiting for a ewe to lamb, he would tie a string from one of her hind legs to his wrist. When her labor pains came and she began to shift around, she would tug the string and he would wake up and tend to her.

And so the talent for what he "had to do" was in large measure the ability to bear the good outcome in mind: to envision, in spite of rocks and gullies, the good health of the fields; to foresee in the pregnant ewes and the advancing seasons a good crop of lambs. And it was the ability to carry in his head for nearly half a century the ideal character and pattern of the Southdown, and to measure his animals relentlessly against it—an ability, rare enough, that marked him as a master stockman.

He told me a story that suggests very well the distinction and the effect of that ability. On one of his trips to the International he competed against a western sheepman who had selected his carload of fifty fat lambs out of ten thousand head.

After the Vinewood carload had won the class, this gentleman came up and asked: "How many did you pick yours from, Mr. Besuden?"

"About seventy-five."

"Well," the western breeder said, "I guess it's better to have the right seventy-five than the wrong ten thousand."

But the ability to recognize the right seventy-five is worthless by itself. Just as necessary is the ability to do the

work and to pay attention. To pay attention, above all—
that is another of the persistent themes of Mr. Besuden's
talk and of his life. He is convinced that paying attention
pays, and this sets him apart from the mechanized "modern" farmers who are pushed to accept more responsibility
than they can properly meet, and to work at freeway
speeds. He wrote in his column of the importance of "little
things done on time." He said that they paid, but he knew
that people did them for more than pay.

He told me also about a farmer who wouldn't scrape the
manure off his shoes until he came to a spot that was bare
of grass. "That's what I mean," he said. "You have to keep
it on your mind."

22

New Roots for Agricultural Research

(1981)

Wes Jackson's recently published book, *New Roots For Agriculture,* is a landmark. For some time before the book came out, I had been hearing of it and of its author by way of highly complimentary rumor, and the book did not disappoint me. It offers a sound, thoroughly documented criticism of the assumptions and the effects of industrial agriculture; for that alone the book would be valuable. But it goes beyond criticism to propose practical remedies, preeminent among them the idea of developing perennial grain crops, or as he calls them, "herbaceous perennial seed-producing polycultures."

What he is proposing, in other words, is a grain field that would lie under the same live vegetative cover year after year, like a pasture. And, like a good pasture, it would not be seeded to a monoculture, but to a mixture of plants, not only to increase productivity, but to increase the range of nutritive value, to reduce the dependence on purchased nitrogen, to reduce vulnerability to pests and diseases—in short, to benefit in every possible way from the principle of diversity.

One does not need to reflect long upon the worst problems and weaknesses of our present agriculture in order to see the significance of this possibility. Perennial grains, once the plantings were established, would be an ideal remedy for soil erosion. Not only would our currently disastrous soil losses be prevented but the soil would build and heal under the continuous cover, exactly as it does under well-managed pasture. The dependence on irrigation would be reduced, for these crops would use water more efficiently than annual monocultures, and the ability of the soil to absorb and hold water would be greatly increased. There would also be large reductions of the use of toxic chemicals and chemical fertilizers and of the dependence on fossil fuels, for after the initial planting the only field work would be that of harvesting.

Wes Jackson is an extraordinary man both in the quality of his insight and in his forthright assumption that radical problems call for radical solutions. To read his book is to wonder what lies behind it, and—because it is so intent upon certain possibilities—what may lie ahead of it in the plans and expectations of its author.

What lies behind it, of course, is a story. The book was written by Wes Jackson, but the life and work it comes out of belong both to Wes and to his wife, Dana. Both are native Kansans of rural background. One of Dana's grandfathers drove cattle up the Chisholm Trail to Abilene. Her father was born in a "dugout." Her parents grew up on farms, but later moved to the small town where she grew up. Both of Wes's grandfathers came to Kansas when it was cattle country. One of them later farmed a quarter section in the Kansas River Valley; the other was in the nursery business. Wes grew up "on the end of a hoe handle," helping his parents, who had a diversified farm on which they grew both grain and truck crops.

The Jacksons met when Dana was a senior in high school and Wes was beginning college. After graduating from Kansas Wesleyan University at Salina, where Wes majored in biology and Dana in English, they married and started a family. Wes earned an M.A. in botany at the University of Kansas, and then after a stint of high school and college teaching they went on to North Carolina State University at Raleigh, where Wes studied for a Ph.D. in genetics. That work completed, they returned to Kansas, and Wes taught at Kansas Wesleyan from 1967 to 1971. In 1971, they went to California, where Wes had taken a job at Sacramento State University. So far it looks like the upward and outward story of a typical academic career. But that wasn't what it was going to turn out to be. By the time they left for California, forces were already in play that would take them back to Kansas and out of academic life.

Dana and Wes think of 1967, after Wes's emergence from the preoccupations of graduate school, as the time when they began to "grow together." It was a disturbing time. They were young parents, much aware of their involvement in the world; they could not ignore the struggle over civil rights and the Vietnam War, the spreading debate about the status of women, the threat of population growth, the public questioning and apparent decline of marriage and family life. And as Kansans, whose family memories spanned the history of the state from the cattle drives to the agri-industry of wheat, they were perhaps peculiarly vulnerable to the deepening worries about soil loss, pollution, resource depletion, etc. They talked about these problems and read about them together, and they began to make some connections. Political, social, and ecological deterioration, they thought, mirrored an inward condition; the outward problems connected in spiritual blight. For that reason, technological solutions alone would solve nothing.

In their own lives the time also was what Wes calls "the age of land hunger." They were looking for a place—or, rather, for *the* place that would be the right spot for them. As if from the homesteaders behind them had come the desire for "the satisfaction of full title and freedom from debt." The place, when they found it, was three acres on the Smoky Hill River just outside of Salina; it is still their home, the three acres having increased by now to twenty-eight.

They bought the place and built a house on it in 1971 — just before, still uncertain of their purpose, they went to California. One thing that adventure proved was that the place they had thought was the right spot *was* the right spot: California "just wasn't the place to be." They were certain now that, as Kansans, they belonged in Kansas. They could be happiest and most useful at home, where they knew the talk "in sale barn or capitol." To remain in California would be to live with too many unanswered questions about their place and their work. And so in 1974 they came home to the house they had built four years earlier but had never lived in.

Dana and Wes spent the next two years "fooling around, trying to figure out a way to live here." The idea they kept working at was the idea of a school. They had some ideas about what was wrong and what to do about it, and they had some ideas about education. Why not take on a few students? Their friend John Simpson, a neighboring attorney, politician, and environmentalist, helped them raise the money to get started. Thus began The Land Institute, "a non-profit educational research organization, devoted to the study of sustainable alternatives in agriculture, energy, waste-management, and shelter."

And then, just at the beginning, on October 17, 1976, their school building burned. They lost tools, books, everything. In their two years of "fooling around," Wes and Dana had been working together on a book, *Toward an Ecological Ethic.* Draft copies survived the fire, but their working copy, notes, and references all burned, and the book was never published.

But the thinking they had worked out in their book was the thinking that had defined their school. The school existed, burned or not: "We didn't think books and tools were what it was all about." The school, as the fire proved, had friends. People sent books and other necessary things. The Jacksons built back, and went ahead.

Now, in addition to the house and the school building, there is a barn with the other outbuildings necessary to subsistence farming; there are storage sheds, a workshop, a greenhouse, solar collectors, windmills. There are gardens and test plots. There is a prairie herbary, containing "perennial native and naturalized grasses and wild flowers of the prairie states," which is to be used for teaching and research, but is also "a savings bank" of native species. The one place is home and farm, campus, experiment station, laboratory, and museum. It is a place to live and work, to teach and learn.

Eight students per semester, they thought, would be the right number. But that has turned out to mean a limit of eight *new* students per semester, for some usually stay over for another term; in the spring of 1981, for example, there will be twelve students, of whom five will be new. But much depends on keeping the number small. That way you don't have the exclusive categories of teachers and students. In so small a group, in which everybody knows everybody else, people tend to deal with each other in pairs, and in every pair somebody is teaching and somebody is learning. The idea is to make "an intellectual climate for spontaneous collaboration rather than minimal compliance." (What teacher or parent does not know about minimal compliance?) The idea is to think a problem through and understand it as fully as possible, not to learn a ready-made answer; the problems of most interest at The

Land Institute don't *have* ready-made answers. The students are considered "participants in a community that promotes sustainability instead of strength through exhaustion."

Perhaps the most important of the Jackson's educational principles is their repudiation of the "professional objectivity" and the ethical indifference of academic specialization. "Knowledge," Wes Jackson says, "is *not* value free." And he says that "the purpose of education is the transmission of values." The Land Institute, therefore, operates not only according to carefully specified purposes, but on some clearly settled and emphatically stated assumptions: that, for example, nuclear power should be opposed; that women should not be exploited; that no interest or value should be put above the health of the land. And these and other such assumptions amount, in effect, to a rather stringent entrance requirement: "If not interested, don't come."

The curriculum does not consist of "courses" but of a set of steps repeated every semester:

1. Development of a checklist of environmental problems.
2. Attempts to analyze and understand the problems.
3. A search for new or alternative solutions.
4. Work toward an ecological ethic.

A student, at step three, may work on a problem of energy, shelter, or waste management, but the school's primary focus is on the problem of soil loss and the development of a sustainable agriculture.

In this work, Wes has been guided by the assumption that you cannot understand agriculture on its own terms. Agriculture depends on nature and is contained in nature; if you want to understand agriculture, therefore, you must understand what preceded it. This is an insight characteristic of those farmers and students of farming whose orientation may be said to be *organic*. Sir Albert Howard approached agriculture by way of an understanding of the self-sustaining ecology of the forest. As a Kansan, Wes Jackson has had to proceed by way of the study of the native ecosystem of the prairie. Looked at on their own terms, present day grainfields can only produce a kind of bewilderment: their productivity is undeniably great, but the costs of production in soil and energy are even greater. More usable energy is going into those fields than is coming out of them, and the tonnage of soil lost to erosion is apt to be several times greater than that of the harvested grain. How come? And what to do?

The characteristic "agribusiness" response is to tinker with the available industrial technology and to try to decide how much ecological loss is "acceptable." But, as Wes Jackson has been at pains to demonstrate, the situation is not *that* desperate: there is something else you can do. You can, so to speak, put a cornfield beside a plot of the native prairie (of which some few patches fortunately still survive), and you can ask another question: What is the difference? The most noticeable difference is that whereas the soil is washing away in the cornfield, it is *building* in the prairie. And there is another difference that explains that one: the corn is an annual, the cornfield is an annual monoculture, but the dominant feature of the native prairie sod is that it is composed of a balanced *diversity* of perennials: grasses, legumes, sunflowers, etc., etc. The prairie is self-renewing; it accumulates ecological capital; and by its own abounding fertility and diversity it controls pests and diseases. The "agribusiness" cornfield, on the other hand, is self-destructive; it consumes more ecological capital than it produces; and, because it is a monoculture, it *invites* pests and diseases.

Wes Jackson's criticism of agriculture grows out of his

understanding of those differences, an understanding that is at once biological, agricultural, and cultural. His research is founded on the paradigm or example of the prairie. If you want to stop soil erosion (so the prairie shows) you have to keep the ground covered—all the time, winter and summer. If you want to keep the ground covered all the time, the best way to do it is with a diverse, mutually beneficial polyculture of perennial plants.

Where are you going to get the perennial plants? There are two possibilities, Wes says: you can breed perennial strains of traditional crops, or you can breed more productive strains of native seed-producing perennials. He would prefer not to use the traditional crops, he says, because they have a bred-in dependence on several things that are wrong with our present way of grain farming: monoculture, cultivation, heavy use of chemicals and fossil fuel. For that reason, and because his aim is polyculture, he has chosen to start from scratch with the wild native species, and to work according to a more complex genetic standard, asking not just how much a plant will yield, but also how well it will fit in with other desirable species.

To be as self-sustaining as possible, he thinks, the mixture of plants will have to contain at least twenty-five percent legumes. And as a minimum desirable yield he has set 1800 lbs. per acre—the weight of thirty bushels of wheat.

Well, then, what are the chances of success? It's a fine idea, but is it possible? "Given a little bit of money and up to a hundred years," Wes says, "we can do it. I see no reason why it can't be done." Perhaps it can be done in less time. He hopes so, but to set an earlier deadline is to risk underestimating the difficulty. It won't be easy: "Are herbaceous perennialism and high yield mutually exclusive? I expect that to be the hardest question put to me. But already,

without selection, we have some very strong possibilities."
A native grass, Sand Dropseed *(Sporobolus cryptandrus)*,
for example, has produced an annual seed yield of 900 lbs.
per acre; Illinois Bundle Flower *(Desmanthus illinoensis)*,
with irrigation, has yielded 1,189 lbs.; Greyheaded Cone-
flower *(Ratibida pinnata)* has yielded 1,600; and Maximil-
lian Sunflower *(Helianthus maximilliani)* has yielded
1,300.

Wes is encouraged by the success in the breeding of corn
in the last fifty or so years: "During the 1920s, corn yielded
around thirty bushels per acre. The 1979 crop averaged 106
bushels per acre . . ." (Part of that increase was caused by
the expanding use of commercial fertilizer, but a considerable amount of credit belongs to the plant breeders.) He
thinks that "the prospects for yield improvement in the perennial may ultimately be as high as for the annual." And
certain biological principles favor the work with perennials, for they are generally easier to crossbreed than annuals, and the resulting hybrids are much more likely to be
fertile than are annual hybrids.

One of the advantages of this project is the possibility of
secondary benefits, such as the improvement of the forage
value of certain plants. Eastern gama grass, for instance,
produces as much dry matter as corn, and its seed is three
times higher in protein. Its weaknesses are that its seed
yield is very small, and the seeds have so poor a germina-
tion rate that the plant now has to be vegetatively propa-
gated. If a strain with viable seed could be produced, then
this plant might become useful as forage long before it
could be made usable as a grain crop.

There are worries, of course, The main practical difficulty
is that a perennial, unlike an annual, cannot afford to concentrate all of its energy into seed; some energy must be

saved to maintain its root system through the winter and to start growth again in the spring—thus "the hardest question" concerns a possible conflict between herbaceous perennialism and high yield.

There are moral worries too. The things that are wrong with agriculture now all come from the human willingness to manipulate nature. The scientists have found more and more ways to manipulate nature, and those ways have been used by industrialists to convert health into wealth. And now Wes Jackson is trying to correct the bad results of that by yet another manipulation of nature. What is there to keep his success, if he succeeds, from being taken up by corporations and turned against nature herself and against the human hopes that have been the inspiration of Wes's work?

There is no way to know for sure—if there were, one would not worry—but then, this is not your typical "agribusiness" research project, either. In the first place, it is based on a natural rather than an industrial paradigm: the farm will be an imitation of a prairie, not of a factory. And it will be understood as a home, a "hearth," for people, not as a "workplace" for "labor." There are at least two considerations which suggest that the development of perennial grain crops will favor a return to small farm agriculture: (1) Perennial grain crops will greatly reduce expenditures for machinery, energy, labor, chemicals, irrigation, and seed. Management will again tend to take precedence over technology and capital. And (2) they will permit the safe use of lands now considered marginal because of vulnerability to erosion under present cropping systems; thus they will reduce pressure on the land market and make it possible for more people to buy land.

But an agriculture based on perennial grains, as Wes Jackson well knows, will involve complex economic, social, and cultural changes, perplexing foresight. For now, it is necessary to understand as much as possible, and to work as responsibly as possible in the light of what one has understood. It is necessary to be prepared for one's work to succeed in a hundred years.

Though Wes Jackson is by training and interest a scientist, he is not just another expert casting bits of paper into the wind of "technological progress." He works as a scientist because he is much more than that: a husband, a father, a teacher, a citizen, a native Kansan, a lover of decency and good sense. By refusing to ignore or disown his own character, he has knocked down the barrier between life and work; and his wife, his children, and his students have become his coworkers. Home, farm, laboratory, and school have become the belongings of the same life and place.

What will come of it? Thinking about it, one feels sure that good will come of it. But one feels too that the question may be unnecessary. What Wes Jackson *will* give us may even be a little beside the point, for there is inestimable value and encouragement in what he has already supplied: an example of a thoroughly informed, technically competent, practical intelligence working by the measure of high ecological and cultural standards—a specialist practioner whose questions and criteria are *not* specialized.

23

Seven Amish Farms

(1981)

In typical Midwestern farming country the distances between inhabited houses are stretching out as bigger farmers buy out their smaller neighbors in order to "stay in." The signs of this "movement" and its consequent specialization are everywhere: good houses standing empty, going to ruin; good stock barns going to ruin; pasture fences fallen down or gone; machines too large for available doorways left in the weather; windbreaks and woodlots gone down before the bulldozers; small schoolhouses and churches deserted or filled with grain.

In the latter part of March this country shows little life. Field after field lies under the dead stalks of last year's corn and soybeans, or lies broken for the next crop; one may drive many miles between fields that are either sodded or planted in winter grain. If the weather is wet, the country will seem virtually deserted. If the ground is dry enough to support their wheels, there will be tractors at work, huge machines with glassed cabs, rolling into the distances of fields larger than whole farms used to be, as solitary as seaborne ships.

The difference between such country and the Amish farmlands in northeast Indiana seems almost as great as that between a desert and an oasis. And it is the *same* difference. In the Amish country there is a great deal more life: more natural life, more agricultural life, more human life. Because the farms are small—most of them containing well under a hundred acres—the Amish neighborhoods are more thickly populated than most rural areas, and you see more people at work. And because the Amish are diversified farmers, their plowed croplands are interspersed with pastures and hayfields and often with woodlots. It is a varied, interesting, healthy looking farm country, pleasant to drive through. When we were there, on the twentieth and twenty-first of last March, the spring plowing had just started, and so you could still see everywhere the annual covering of stable manure on the fields, and the teams of Belgians or Percherons still coming out from the barns with loaded spreaders.

Our host, those days, was William J. Yoder, a widely respected breeder of Belgian horses, an able farmer and carpenter, and a most generous and enjoyable companion. He is a vigorous man, strenuously involved in the work of his farm and in the life of his family and community. From the look of him and the look of his place, you know that he has not just done a lot of work in his time, but has done it well, learned from it, mastered the necessary disciplines. He speaks with heavy stress on certain words—the emphasis of conviction, but also of pleasure, for he enjoys the talk that goes on among people interested in horses and in farming. But unlike many people who enjoy talking, he speaks with care. Bill was born in this community, has lived there all his life, and he has grandchildren who will probably live there all their lives. He belongs there, then, root and branch, and he knows the history and the quality of many of the farms. On the two days, we visited farms belonging to Bill himself, four of his sons, and two of his sons-in-law.

The Amish farms tend to divide up between established

ones, which are prosperous looking and well maintained, and run-down, abused, or neglected ones, on which young farmers are getting started. Young Amish farmers *are* still getting started, in spite of inflation, speculators' prices, and usurious interest rates. My impression is that the proportion of young farmers buying farms is significantly greater among the Amish than among conventional farmers.

Bill Yoder's own eighty-acre farm is among the established ones. I had been there in the fall of 1975 and had not forgotten its aspect of cleanness and good order, its well-kept white buildings, neat lawns, and garden plots. Bill has owned the place for twenty-six years. Before he bought it, it had been rented and row cropped, with the usual result: it was nearly played out. "The buildings," he says, "were nothing," and there were no fences. The first year, the place produced five loads (maybe five tons) of hay, "and that was mostly sorrel." The only healthy plants on it were the spurts of grass and clover that grew out of the previous year's manure piles. The corn crop that first year "might have been thirty bushels an acre," all nubbins. The sandy soil blew in every strong wind, and when he plowed the fields his horses' feet sank into "quicksand potholes" that the share uncovered.

The remedy has been a set of farming practices traditional among the Amish since the seventeenth century: diversification, rotation of crops, use of manure, seeding of legumes. These practices began when the Anabaptist sects were disfranchised in their European homelands and forced to the use of poor soil. We saw them still working to restore farmed-out soils in Indiana. One thing these practices do is build humus in the soil, and humus does several things: increases fertility, improves soil structure, improves both water-holding capacity and drainage. "No humus, you're in trouble," Bill says.

After his rotations were established and the land had

begun to be properly manured, the potholes disappeared, and the soil quit blowing. "There's something in it now—there's some substance there." Now the farm produces abundant crops of corn, oats, wheat, and alfalfa. Oats now yield 90–100 bushels per acre. The corn averages 100–125 bushels per acre, and the ears are long, thick, and well filled.

Bill's rotation begins and ends with alfalfa. Every fall he puts in a new seeding of alfalfa with his wheat; every spring he plows down an old stand of alfalfa, "no matter how good it is." From alfalfa he goes to corn for two years, planting thirty acres, twenty-five for ear corn and five for silage. After the second year of corn, he sows oats in the spring, wheat and alfalfa in the fall. In the fourth year the wheat is harvested; the alfalfa then comes on and remains through the fifth and sixth years. Two cuttings of alfalfa are taken each year. After curing in the field, the hay is hauled to the barn, chopped, and blown into the loft. The third cutting is pastured.

Unlike cow manure, which is heavy and chunky, horse manure is light and breaks up well coming out of the spreader; it interferes less with the growth of small seedlings and is less likely to be picked up by a hay rake. On Bill's place, horse manure is used on the fall seedings of wheat and alfalfa, on the young alfalfa after the wheat harvest, and both years on the established alfalfa stands. The cow manure goes on the corn ground both years. He usually has about 350 eighty-bushel spreader loads of manure, and each year he covers the whole farm—cropland, hayland, and pasture.

With such an abundance of manure there obviously is no *dependence* on chemical fertilizers, but Bill uses some as a "starter" on his corn and oats. On corn he applies 125 pounds of nitrogen in the row. On oats he uses 200–250 pounds of 16-16-16, 20-20-20, or 24-24-24. He routinely spreads two tons of lime to the acre on the ground being prepared for wheat.

His out-of-pocket costs per acre of corn last year were as follows:

Seed (planted at a rate of seven acres per bushel) . . . . . $ 7.00
Fertilizer 7.75
Herbicide (custom applied, first year only) 16.40

That comes to a total of $31.15 per acre—or, if the corn makes only a hundred bushels per acre, a little over $0.31 per bushel. In the second year his per acre cost is $14.75, less than $0.15 per bushel, bringing the two-year average to $22.95 per acre or about $0.23 per bushel.

The herbicide is used because, extra horses being on the farm during the winter, Bill has to buy eighty to a hundred tons of hay, and in that way brings in weed seed. He had no weed problem until he started buying hay. Even though he uses the herbicide, he still cultivates his corn three times.

His cost per acre of oats came to $33.00 ($12.00 for seed and $21.00 for fertilizer)—or, at ninety bushels per acre, about $0.37 per bushel.

Of Bill's eighty acres, sixty-two are tillable. He has ten acres of permanent pasture, and seven or eight of woodland, which produced the lumber for all the building he has done on the place. In addition, for $500 a year he rents an adjoining eighty acres of "hill and woods pasture" which provides summer grazing for twenty heifers; and on another neighboring farm he rents varying amounts of cropland.

All the field work is done with horses, and this, of course, comes virtually free—a by-product of the horse-breeding enterprise. Bill has an ancient Model D John Deere tractor that he uses for belt power.

At the time of our visit, there were twenty-two head of horses on the place. But that number was unusually low, for Bill aims to keep "around thirty head." He has a band of excellent brood mares and three stallions, plus young stock of assorted ages. Since October 1 of last year, he had sold eighteen head of registered Belgian horses. In the winters he operates a "urine line," collecting "pregnant mare urine," which is sold to a pharmaceutical company for the extraction of various hormones. For this purpose he boards a good many mares belonging to neighbors; that is why he must buy the extra hay that causes his weed problem. (Horses are so numerous on this farm because they are one of its money-making enterprises. If horses were used only for work on this farm, four good geldings would be enough.)

One bad result of the dramatic rise in draft horse prices over the last eight or ten years is that it has tended to focus attention on such characteristics as size and color to the neglect of less obvious qualities such as good feet. To me, foot quality seems a critical issue. A good horse with bad feet is good for nothing but decoration, and at sales and shows there are far too many flawed feet disguised by plastic wood and black shoe polish. And so I was pleased to see that every horse on Bill Yoder's place had sound, strong-walled, correctly shaped feet. They were good horses all around, but their other qualities were well founded; they stood on good feet, and this speaks of the thoroughness of his judgment and also of his honesty.

Though he is a master horseman, and the draft horse business is more lucrative now than ever in its history, Bill does not specialize in horses, and that is perhaps the clearest indication of his integrity as a farmer. Whatever may be the dependability of the horse economy, on this farm it rests upon a diversified agricultural economy that is sound.

He was milking five Holstein cows; he had fifteen Holstein heifers that he had raised to sell; and he had just marketed thirty finished hogs, which is the number that he usually has on fence. All the animals had been well wintered—Bill quotes his father approvingly: "Well wintered is half summered"—and were in excellent condition. Another saying of his father's that Bill likes to quote—"Keep the horses on the side of the fence the feed is on"—has obviously been obeyed here. The feeding is careful, the feed is good, and it is abundant. Though it was almost spring, there were ample surpluses in the hayloft and in the corn cribs.

Other signs of the farm's good health were three sizable garden plots, and newly pruned grapevines and raspberry canes. The gardener of the family is Mrs. Yoder. Though most of the children are now gone from home, Bill says that she still grows as much garden stuff as she ever did.

All seven of the Yoders' sons live in the community. Floyd, the youngest, is still at home. Harley has a house on nearly three acres, works in town, and returns in the afternoons to his own shop where he works as a farrier. Henry, who also works in town, lives with Harley and his wife. The other four sons are now settled on farms that they are in the process of paying for. Richard has eighty acres, Orla eighty, Mel fifty-seven, and Wilbur eighty. Two sons in law also living in the community are Perry Bontrager, who owns ninety-five acres, and Ervin Mast, who owns sixty-five. Counting Bill's eighty acres, the seven families are living on 537 acres. Of the seven farms, only Mel's is entirely tillable, the acreages in woods or permanent pasture varying from five to twenty-six.

These young men have all taken over run-down farms, on which they are establishing rotations and soil husbandry practices that, being traditional, more or less resemble

Bill's. It seemed generally agreed that after three years of this treatment the land would grow corn, as Perry Bontrager said, "like anywhere else."

These are good farmers, capable of the intelligent planning, sound judgment, and hard work that good farming requires. Abused land heals and flourishes in their care. None of them expressed a wish to own more land; all, I believe, feel that what they have will be enough—when it is paid for. The big problems are high land prices and high interest rates, the latter apparently being the worst.

The answer, for Bill's sons so far, has been town work. All of them, after leaving home, have worked for Redman Industries, a manufacturer of mobile homes in Topeka. They do piecework, starting at seven in the morning and quitting at two in the afternoon, using the rest of the day for farming or other work. This, Bill thinks, is now "the only way" to get started farming. Even so, there is "a lot of debt" in the community—"more than ever."

With a start in factory work, with family help, with government and bank loans, with extraordinary industry and perseverance, with highly developed farming skills, it is still possible for young Amish families to own a small farm that will eventually support them. But there is more strain in that effort now than there used to be, and more than there should be. When the burden of usurious interest becomes too great, these young men are finding it necessary to make temporary returns to their town jobs.

The only one who spoke of his income was Mel, who owns fifty-seven acres, which, he says, *will be* enough. He and his family milk six Holsteins. He had nine mares on the urine line last winter, seven of which belonged to him. And he had twelve brood sows. Last year his gross income was $43,000. Of this, $12,000 came from hogs, $7,000 from his milk cows, the rest from his horses and the sale of his wheat. After his production costs, but *before* payment of interest, he netted $22,000. In order to cope with the interest payments, Mel was preparing to return to work in town.

These little Amish farms thus become the measure both of "conventional" American agriculture and of the cultural meaning of the national industrial economy.

To begin with, these farms give the lie direct to that false god of "agribusiness": the so-called economy of scale. The small farm is not an anachronism, is not unproductive, is not unprofitable. Among the Amish, it is still thriving, and is still the economic foundation of what John A. Hostetler (in *Amish Society,* third edition) rightly calls "a healthy culture." Though they do not produce the "record-breaking yields" so touted by the "agribusiness" establishment, these farms are nevertheless highly productive. And if they are not likely to make their owners rich (never an Amish goal), they can certainly be said to be sufficiently profitable. The economy of scale has helped corporations and banks, not farmers and farm communities. It has been an economy of dispossession and waste—plutocratic, if not in aim, then certainly in result.

What these Amish farms suggest, on the contrary, is that in farming there is inevitably a scale that is suitable both to the productive capacity of the land and to the abilities of the farmer; and that agricultural problems are to be properly solved, not in expansion, but in management, diversity, balance, order, responsible maintenance, good character, and in the sensible limitation of investment and overhead. (Bill makes a careful distinction between "healthy" and "unhealthy" debt, a "healthy debt" being "one you can hope to pay off in a reasonable way.")

Most significant, perhaps, is that while conventional agriculture, blindly following the tendency of any industry to exhaust its sources, has made soil erosion a national catastrophe, these Amish farms conserve the land and improve it in use.

And what is one to think of a national economy that drives such obviously able and valuable farmers to factory work? What value does such an economy impose upon thrift, effort, skill, good husbandry, family and community health?

In spite of the unrelenting destructiveness of the larger economy, the Amish—as Hostetler points out with acknowledged surprise and respect—have almost doubled in population in the last twenty years. The doubling of a population is, of course, no significant achievement. What is significant is that these agricultural communities have doubled their population *and yet remained agricultural communities* during a time when conventional farmers have failed by the millions. This alone would seem to call for a careful look at Amish ways of farming. That those ways have, during the same time, been ignored by the colleges and the agencies of agriculture must rank as a prime intellectual wonder.

Amish farming has been so ignored, I think, because it involves a complicated structure that is at once biological and cultural, rather than industrial or economic. I suspect that anyone who might attempt an accounting of the economy of an Amish farm would soon find himself dealing with virtually unaccountable values, expenses, and benefits. He would be dealing with biological forces and processes not always measurable, with spiritual and community values not quantifiable; at certain points he would be dealing with mysteries—and he would be finding that these unaccountables and inscrutables have results, among others, that are economic. Hardly an appropriate study for the "science" of agricultural economics.

The economy of conventional agriculture or "agribusiness" is remarkable for the simplicity of its arithmetic. It involves a manipulation of quantities that are all entirely accountable. List your costs (land, equipment, fuel, fertilizer, pesticides, herbicides, wages), add them up, subtract them from your earnings, or subtract your earnings from them, and you have the result.

Suppose, on the other hand, that you have an eighty-acre farm that is not a "food factory" but your home, your given portion of Creation which you are morally and spiritually obliged "to dress and to keep." Suppose you farm, not for wealth, but to maintain the integrity and the practical supports of your family and community. Suppose that, the farm being small enough, you farm it with family work and work exchanged with neighbors. Suppose you have six Belgian brood mares that you use for field work. Suppose that you also have milk cows and hogs, and that you raise a variety of grain and hay crops in rotation. What happens to your accounting then?

To start with, several of the costs of conventional farming are greatly diminished or done away with. Equipment, fertilizer, chemicals all cost much less. Fuel becomes feed, but you have the mares and are feeding them anyway; the work ration for a brood mare is not a lot more costly than a maintenance ration. And the horses, like the rest of the livestock, are making manure. Figure that in, and figure, if you can, the value of the difference between manure and chemical fertilizer. You can probably get an estimate of the value of the nitrogen fixed by your alfalfa, but how will you quantify the value to the soil of its residues and deep roots? Try to compute the value of humus in the soil—in improved drainage, improved drought resistance, improved tilth, improved health. Wages, if you pay your children, will still be among your costs. But compute the difference between paying your children and paying "labor." Work exchanged with neighbors can be reduced to "man-hours" and assigned a dollar value. But compute the difference between a neighbor and "labor." Compute the value of a family or a community to any one of its members. We may, as we must, grant that among the values of family and community there is economic value—but what is it?

In the Louisville *Courier-Journal* of April 5, 1981, the Mobil Oil Corporation ran an advertisement which was yet another celebration of "scientific agriculture." American farming, the Mobil people are of course happy to say, "requires <u>more petroleum products than almost any other industry.</u> A gallon of gasoline to produce a single bushel of corn, for example. . . ." This, they say, enables "each American farmer to feed sixty-seven people." And they say that this is "a-maizing."

Well, it certainly is! And the chances are good that an agriculture totally dependent on the petroleum industry is not yet as amazing as it is going to be. But one thing that is already sufficiently amazing is that a bushel of corn produced by the burning of one gallon of gasoline has already cost more than *six times* as much as a bushel of corn grown by Bill Yoder. How does Bill Yoder escape what may justly be called the petroleum tax on agriculture? He does so by a series of substitutions: of horses for tractors, of feed for fuel, of manure for fertilizer, of sound agricultural methods and patterns for the exploitive methods and patterns of industry. But he has done more than that—or, rather, he and his people and their tradition have done more. They have substituted themselves, their families, and their communities for petroleum. The Amish use little petroleum—and need little—because they have those other things.

I do not think that we can make sense of Amish farming until we see it, until we become willing to see it, as belonging essentially to the Amish practice of Christianity, which instructs that one's neighbors are to be loved as oneself. To farmers who give priority to the maintenance of their community, the economy of scale (that is, the economy of *large* scale, of "growth") can make no sense, for it requires the ruination and displacement of neighbors. A farm cannot be increased except by the decrease of a neighborhood. What the interest of the community proposes is invariably an economy of *proper* scale. A whole set of agricultural proprieties must be observed: of farm size, of methods, of tools, of energy sources, of plant and animal species. Community interest also requires charity, neighborliness, the care and instruction of the young, respect for the old; thus it assures its integrity and survival. Above all, it requires good stewardship of the land, for the community, as the Amish have always understood, is no better than its land. "If treated violently or exploited selfishly," John Hostetler writes, the land "will yield poorly." There could be no better statement of the meaning of the *practice* and the practicality of charity. Except to the insane narrow-mindedness of industrial economics, selfishness does not pay.

The Amish have steadfastly subordinated economic value to the values of religion and community. What is too readily overlooked by a secular, exploitive society is that their ways of doing this are not "empty gestures" and are not "backward." In the first place, these ways have kept the communities intact through many varieties of hard times. In the second place, they conserve the land. In the third place, they yield economic benefits. The community, the religious fellowship, has many kinds of value, and among them is economic value. It is the result of the practice of neighborliness, and of the practice of stewardship.

What moved me most, what I liked best, in those days we spent with Bill Yoder was the sense of the continuity of the community in his dealings with his children and in their dealings with their children.

Bill has helped his sons financially so far as he has been able. He has helped them with his work. He has helped them by sharing what he has—lending a stallion, say, at breeding time, or lending a team. And he helps them by buying good pieces of equipment that come up for sale. "If he ever gets any money," he says of one of the boys, for whom he has bought an implement, "he'll pay me for it. If he don't, he'll just use it." He has been their teacher, and he remains their advisor. But he does not stand before them as a domineering patriarch or "authority figure." He seems to speak, rather, as a representative of family and community experience. In their respect for him, his sons respect their tradition. They are glad for his help, advice, and example, but there is nothing servile in this. It seems to be given and taken in a kind of familial friendship, respect going both ways.

Everywhere we went, when school was not in session, the children were at the barns, helping with the work, watching, listening, learning to farm in the way it is best learned. Wilbur told us that his eleven-year-old son had cultivated twenty-three acres of corn last year with a team and a riding cultivator. That reminded Bill of the way he taught Wilbur to do the same job.

Wilbur was little then, and he loved to sit in his father's lap and drive the team while Bill worked the cultivator. If Wilbur could drive, Bill thought, he could do the rest of it. So he got off and shortened the stirrups so the boy could reach them with his feet. Wilbur started the team, and within a few steps began plowing up the corn.

"Whoa!" he said.

And Bill, who was walking behind him, said, "Come up!"

And it went that way for a little bit:

"Whoa!"

"Come up!"

And then Wilbur started to cry, and Bill said:

"Don't cry! Go ahead!"

V

24

The Gift of Good Land

(1979)

> "Dream not of other Worlds..."
> *Paradise Lost* VIII, 175

My purpose here is double. I want, first, to attempt a Biblical argument for ecological and agricultural responsibility. Second, I want to examine some of the practical implications of such an argument. I am prompted to the first of these tasks partly because of its importance in our unresolved conflict about how we should use the world. That those who affirm the divinity of the Creator should come to the rescue of His creature is a logical consistency of great potential force.

The second task is obviously related to the first, but my motive here is somewhat more personal. I wish to deal directly at last with my own long held belief that Christianity, as usually presented by its organizations, is not *earthly enough*— that a valid spiritual life, in this world, must have a practice and a practicality—it must have a material result. (I am well aware that in this belief I am not alone.) What I shall be working toward is some sort of practical understanding of what Arthur O. Lovejoy called the "this-worldly" aspect of Biblical thought. I want to see if there is not at least implicit in the Judeo-Christian heritage a doctrine such as that the Buddhists call "right livelihood" or "right occupation."

Some of the reluctance to make a forthright Biblical argument against the industrial rape of the natural world seems to come from the suspicion that this rape originates with the Bible, that Christianity cannot cure what, in effect, it has caused. Judging from conversations I have had, the best known spokesman for this view is Professor Lynn White, Jr., whose essay, "The Historical Roots of Our Ecologic Crisis" has been widely published.

Professor White asserts that it is a "Christian axiom that nature has no reason for existence save to serve man." He seems to base his argument on one Biblical passage, Genesis 1:28, in which Adam and Eve are instructed to "subdue" the earth. "Man," says Professor White, "named all the animals, thus establishing his dominance over them." There is no doubt that Adam's superiority over the rest of Creation was represented, if not established, by this act of naming; he *was* given dominance. But that this dominance was meant to be tyrannical, or that "subdue" meant to destroy, is by no means a necessary inference. Indeed, it might be argued that the correct understanding of this "dominance" is given in Genesis 2:15, which says that Adam and Eve were put into the Garden "to dress it and to keep it."

But these early verses of Genesis can give us only limited help. The instruction in Genesis 1:28 was, after all, given to Adam and Eve in the time of their innocence, and it seems certain that the word "subdue" would have had a different intent and sense for them at that time than it could have for them, or for us, after the Fall.

It is tempting to quarrel at length with various statements in Professor White's essay, but he has made that unnecessary by giving us two sentences that define both his problem and my task. He writes, first, that "God planned all of this [the Creation] explicitly for man's benefit and rule: no item in the physical creation had any purpose save to serve man's purposes." And a few sentences later he says: "Christianity . . . insisted that it is God's will that man exploit nature for his *proper* ends" [My emphasis].

It is certainly possible that there might be a critical difference between "man's purposes" and "man's *proper* ends." And one's belief or disbelief in that difference, and one's seriousness about the issue of propriety, will tell a great deal about one's understanding of the Judeo-Christian tradition.

I do not mean to imply that I see no involvement between that tradition and the abuse of nature. I know very well that Christians have not only been often indifferent to such abuse, but have often condoned it and often perpetrated it. That is not the issue. The issue is whether or not the Bible explicitly or implicitly defines a *proper* human use of Creation or the natural world. Proper use, as opposed to improper use, or abuse, is a matter of great complexity, and to find it adequately treated it is necessary to turn to a more complex story than that of Adam and Eve.

The story of the giving of the Promised Land to the Israelites is more serviceable than the story of the giving of the Garden of Eden, because the Promised Land is a divine gift to a *fallen* people. For that reason the giving is more problematical, and the receiving is more conditional and more difficult. In the Bible's long working out of the understanding of this gift, we may find the beginning—and, by implication, the end—of the definition of an ecological discipline.

The effort to make sense of this story involves considerable difficulty because the tribes of Israel, though they see the Promised Land as a gift to them from God, are also obliged to take it by force from its established inhabitants. And so a lot of the "divine sanction" by which they act sounds like the sort of rationalization that invariably accompanies nationalistic aggression and theft. It is impossible to ignore the similarities to the westward movement of the American frontier. The Israelites were following their own doctrine of "manifest destiny," which for them, as for us, disallowed any human standing to their opponents. In Canaan, as in America, the conquerors acted upon the broadest possible definition of idolatry and the narrowest possible definition of justice. They conquered with the same ferocity and with the same genocidal intent.

But for all these similarities, there is a significant difference. Whereas the greed and violence of the American frontier produced an ethic of greed and violence that justified American industrialization, the ferocity of the conquest of Canaan was accompanied from the beginning by the working out of an ethical system antithetical to it—and antithetical, for that matter, to the American conquest with which I have compared it. The difficulty but also the wonder of the story of the Promised Land is that, there, the primordial and still continuing dark story of human rapaciousness begins to be accompanied by a vein of light which, however improbably and uncertainly, still accompanies us. This light originates in the idea of the land as a gift—not a free or a deserved gift, but a gift given upon certain rigorous conditions.

It is a gift because the people who are to possess it did not create it. It is accompanied by careful warnings and demonstrations of the folly of saying that "My power and the might of mine hand hath gotten me this wealth" (Deuteronomy 8:17). Thus, deeply implicated in the very definition of this gift is a specific warning against *hubris* which is the great ecological sin, just as it is the great sin of politics. People are not gods. They must not act like gods or assume godly authority. If they do, terrible retributions are in store. In this warning we have the root of the idea of propriety, of *proper* human purposes and ends. We must not use the world as though we created it ourselves.

The Promised Land is not a permanent gift. It is "given,"

but only for a time, and only for so long as it is properly used. It is stated unequivocally, and repeated again and again, that "the heaven and the heaven of heavens is the Lord's thy God, the earth also, with all that therein is" (Deuteronomy 10:14). What is given is not ownership, but a sort of tenancy, the right of habitation and use: "The land shall not be sold forever: for the land is mine; for ye are strangers and sojourners with me" (Leviticus 25:23).

In token of His landlordship, God required a sabbath for the land, which was to be left fallow every seventh year; and a sabbath of sabbaths every fiftieth year, a "year of jubilee," during which not only would the fields lie fallow, but the land would be returned to its original owners, as if to free it of the taint of trade and the conceit of human ownership. But beyond their agricultural and social intent, these sabbaths ritualize an observance of the limits of "my power and the might of mine hand"—the limits of human control. Looking at their fallowed fields, the people are to be reminded that the land is theirs only by gift; it exists in its own right, and does not begin or end with any human purpose.

The Promised Land, moreover, is "a land which the Lord thy God careth for: the eyes of the Lord thy God are always upon it" (Deuteronomy 11:12). And this care promises a repossession by the true landlord, and a fulfillment not in the power of its human inhabitants: "as truly as I live, all the earth shall be filled with the glory of the Lord" (Numbers 14:21)—a promise recalled by St. Paul in Romans 8:21: "the creature [the Creation] itself also shall be delivered from the bondage of corruption into the glorious liberty of the children of God."

Finally, and most difficult, the good land is not given as a reward. It is made clear that the people chosen for this gift do not deserve it, for they are "a stiff-necked people" who have been wicked and faithless. To such a people such a gift can be given only as a moral predicament: having failed to deserve it beforehand, they must prove worthy of it afterwards; they must use it well, or they will not continue long in it.

How are they to prove worthy?

First of all, they must be faithful, grateful, and humble; they must remember that the land is a gift: "When thou hast eaten and art full, then thou shalt bless the Lord thy God for the good land which he hath given thee." (Deuteronomy 8:10).

Second, they must be neighborly. They must be just, kind to one another, generous to strangers, honest in trading, etc. These are social virtues, but, as they invariably do, they have ecological and agricultural implications. For the land is described as an "inheritance"; the community is understood to exist not just in space, but also in time. One lives in the neighborhood, not just of those who now live "next door," but of the dead who have bequeathed the land to the living, and of the unborn to whom the living will in turn bequeath it. But we can have no direct behavioral connection to those who are not yet alive. The only neighborly thing we can do for them is to preserve their inheritance: we must take care, among other things, of the land, which is never a possession, but an inheritance to the living, as it will be to the unborn.

And so the third thing the possessors of the land must do to be worthy of it is to practice good husbandry. The story of the Promised Land has a good deal to say on this subject, and yet its account is rather fragmentary. We must depend heavily on implication. For sake of brevity, let us consider just two verses (Deuteronomy 22:6–7):

> If a bird's nest chance to be before thee in the way in any tree, or on the ground, whether they be young ones, or eggs, and the dam sitting upon the young, or upon the eggs, thou shalt not take the dam with the young:

But thou shalt in any wise let the dam go, and take the young to thee; that it may be well with thee, and that thou mayest prolong thy days.

This, obviously, is a perfect paradigm of ecological and agricultural discipline, in which the idea of inheritance is necessarily paramount. The inflexible rule is that the source must be preserved. You may take the young, but you must save the breeding stock. You may eat the harvest, but you must save seed, and you must preserve the fertility of the fields.

What we are talking about is an elaborate understanding of charity. It is so elaborate because of the perception, implicit here, explicit in the New Testament, that charity by its nature cannot be selective—that it is, so to speak, out of human control. It cannot be selective because between any two humans, or any two creatures, all Creation exists as a bond. Charity cannot be just human, any more than it can be just Jewish or just Samaritan. Once begun, wherever it begins, it cannot stop until it includes all Creation, for all creatures are parts of a whole upon which each is dependent, and it is a contradiction to love your neighbor and despise the great inheritance on which his life depends. Charity even for one person does not make sense except in terms of an effort to love all Creation in response to the Creator's love for it.

And how is this charity answerable to "man's purposes"? It is not, any more than the Creation itself is. Professor White's contention that the Bible proposes any such thing is, so far as I can see, simply wrong. It is not allowable to love the Creation according to the purposes one has for it, any more than it is allowable to love one's neighbor in order to borrow his tools. The wild ass and the unicorn are said in the Book of Job (39:5–12) to be "free," precisely in the sense that they are not subject or serviceable to human purposes. The same point—though it is not the main point of that passage—is made in the Sermon on the Mount in reference to "the fowls of the air" and "the lilies of the field." Faced with this problem in Book VIII of *Paradise Lost,* Milton scrupulously observes the same reticence. Adam asks about "celestial Motions," and Raphael refuses to explain, making the ultimate mysteriousness of Creation a test of intellectual propriety and humility:

> . . . for the Heav'n's wide Circuit, let it speak
> The Maker's high magnificence, who built
> So spacious, and his Line stretcht out so far;
> That Man may know he dwells not in his own;
> An Edifice too large for him to fill,
> Lodg'd in a small partition, and the rest
> Ordain'd for uses to his Lord best known.
>
> *(lines 100–106)*

The Creator's love for the Creation is mysteriously precisely because it does not conform to human purposes. The wild ass and the wild lilies are loved by God for their own sake and yet they are part of a pattern that we must love because it includes us. This is a pattern that humans can understand well enough to respect and preserve, though they cannot "control" it or hope to understand it completely. The mysterious and the practical, the Heavenly and the earthly, are thus joined. Charity is a theological virtue and is prompted, no doubt, by a theological emotion, but it is also a practical virtue because it must be practiced. The requirements of this complex charity cannot be fulfilled by smiling in abstract beneficence on our neighbors and on the scenery. It must come to acts, which must come from skills. Real charity calls for the study of agriculture, soil husbandry, engineering, architecture, mining, manufacturing, transportation, the making of monuments and pictures, songs and stories. It calls not just for skills but for the study and criticism of skills, because in all of them a choice must be made: they can be used either charitably or uncharitably.

How can you love your neighbor if you don't know how to build or mend a fence, how to keep your filth out of his water supply and your poison out of his air; or if you do not produce anything and so have nothing to offer, or do not take care of yourself and so become a burden? How can you be a neighbor without *applying* principle—without bringing virtue to a practical issue? How will you practice virtue without skill?

The ability to be good is not the ability to do nothing. It is not negative or passive. It is the ability to do something well—to do good work for good reasons. In order to be good you have to know how—and this knowing is vast, complex, humble and humbling; it is of the mind and of the hands, of neither alone.

The divine mandate to use the world justly and charitably, then, defines every person's moral predicament as that of a steward. But this predicament is hopeless and meaningless unless it produces an appropriate discipline: stewardship. And stewardship is hopeless and meaningless unless it involves long-term courage, perseverance, devotion, and skill. This skill is not to be confused with any accomplishment or grace of spirit or of intellect. It has to do with everyday proprieties in the practical use and care of created things—with "right livelihood."

If "the earth is the Lord's" and we are His stewards, then obviously some livelihoods are "right" and some are not. Is there, for instance, any such thing as a Christian strip mine? A Christian atomic bomb? A Christian nuclear power plant or radioactive waste dump? What might be the design of a Christian transportation or sewer system? Does not Christianity imply limitations on the scale of technology, architecture, and land holding? Is it Christian to profit or otherwise benefit from violence? Is there not, in Christian ethics, an implied requirement of practical separation from a destructive or wasteful economy? Do not Christian values require the enactment of a distinction between an organization and a community?

It is impossible to understand, much less to answer, such questions except in reference to issues of practical skill, because they all have to do with distinctions between kinds of action. These questions, moreover, are intransigently personal, for they ask, ultimately, how each livelihood and each life will be taken from the world, and what each will cost in terms of the livelihoods and lives of others. Organizations and even communities cannot hope to answer such questions until individuals have begun to answer them.

But here we must acknowledge one inadequacy of Judeo-Christian tradition. At least in its most prominent and best known examples, this tradition does not provide us with a precise enough understanding of the commonplace issues of livelihood. There are two reasons for this.

One is the "otherworldly philosophy" that, according to Lovejoy, "has, in one form or another, been the dominant official philosophy of the larger part of civilized mankind through most of its history. . . . The greater number of the subtler speculative minds and of the great religious teachers have . . . been engaged in weaning man's thought or his affections, or both, from . . . Nature." The connection here is plain.

The second reason is that the Judeo-Christian tradition as we have it in its art and literature, including the Bible, is so strongly heroic. The poets and storytellers in this tradition have tended to be interested in the extraordinary actions of "great men"—actions unique in grandeur, such as may occur only once in the history of the world. These extraordinary actions do indeed bear a universal significance, but they cannot very well serve as examples of ordinary behavior. Ordinary behavior belongs to a different dramatic mode, a different understanding of action, even a different understanding of virtue. The drama of heroism raises above all the issue of physical and moral courage: Does the hero have, in extreme circumstances, the courage to obey—to perform the task, the sacrifice, the resistance, the pilgrimage that he is called on to perform? The drama of ordinary or daily behavior also raises the issue of courage, but it raises at the same time the issue of skill; and, because ordinary behavior lasts so much longer than heroic action, it raises in a more complex and difficult way the issue of perseverance. It may, in some ways, be easier to be Samson than to be a good husband or wife day after day for fifty years.

These heroic works are meant to be (among other things) instructive and inspiring to ordinary people in ordinary life, and they are, grandly and deeply so. But there are two issues that they are prohibited by their nature from raising: the issue of life-long devotion and perseverance in unheroic tasks, and the issue of good workmanship or "right livelihood."

It can be argued, I believe, that until fairly recently there was simply no need for attention to such issues, for there existed yeoman or peasant or artisan classes: these were the people who did the work of feeding and clothing and housing, and who were responsible for the necessary skills, disciplines, and restraints. As long as those earth-keeping classes and their traditions were strong, there was at least the hope that the world would be well used. But probably the most revolutionary accomplishment of the industrial revolution was to destroy the traditional livelihoods and so break down the cultural lineage of those classes.

The industrial revolution has held in contempt not only the "obsolete skills" of those classes, but the concern for quality, for responsible workmanship and good work, that supported their skills. For the principle of good work it substituted a secularized version of the heroic tradition: the ambition to be a "pioneer" of science or technology, to make a "breakthrough" that will "save the world" from some "crisis" (which now is usually the result of some previous "breakthrough").

The best example we have of this kind of hero, I am afraid, is the fallen Satan of *Paradise Lost*—Milton undoubtedly having observed in his time the prototypes of industrial heroism. This is a hero who instigates and influences the actions of others, but does not act himself. His heroism is of the mind only—escaped as far as possible, not only from divine rule, from its place in the order of creation or the Chain of Being, but also from the influence of material creation:

> A mind not to be chang'd by Place or Time.
> The mind is its own place, and in itself
> Can make a Heav'n of Hell, a Hell of Heav'n.
> *(Book I, lines 253–255)*

This would-be heroism is guilty of two evils that are prerequisite to its very identity: *hubris* and abstraction. The industrial hero supposes that "mine own *mind* hath saved me"—and moreover that it may save the world. Implicit in this is the assumption that one's mind is one's own, and that it may choose its own place in the order of things; one usurps divine authority, and thus, in classic style, becomes the author of results that one can neither foresee nor control.

And because this mind is understood only as a cause, its primary works are necessarily abstract. We should remind ourselves that materialism in the sense of the love of material things is not in itself an evil. As C. S. Lewis pointed out, God too loves material things; He invented them. The Devil's work is abstraction—not the love of material things, but the love of their quantities—which, of course, is why "David's heart smote him after that he had numbered the people" (II Samuel 24:10). It is not the lover of material

things but the abstractionist who defends long-term damage for short-term gain, or who calculates the "acceptabil-
ity" of industrial damage to ecological or human health, or
who counts dead bodies on the battlefield. The true lover
of material things does not think in this way, but is answerable instead to the paradox of the parable of the lost sheep:
that each is more precious than all.

But perhaps we cannot understand this secular heroic
mind until we understand its opposite: the mind obedient
and in place. And for that we can look again at Raphael's
warning in Book VIII of *Paradise Lost:*

> . . . apt the Mind or Fancy is to rove
> Uncheckt, and of her roving is no end;
> Till warn'd, or by experience taught, she learn
> That not to know at large of things remote
> From use, obscure and subtle, but to know
> That which before us lies in daily life,
> Is the prime Wisdom; what is more, is fume,
> Or emptiness, or fond impertinence,
> And renders us in things that most concern
> Unpractic'd, unprepar'd, and still to seek.
> Therefore from this high pitch let us descend
> A lower flight, and speak of things at hand
> Useful . . .
>
> *(lines 188–200)*

In its immediate sense this is a warning against thought
that is theoretical or speculative (and therefore abstract),
but in its broader sense it is a warning against disobedience—the eating of the forbidden fruit, an act of *hubris,*
which Satan justifies by a compellingly reasonable theory
and which Eve undertakes as a speculation.

A typical example of the conduct of industrial heroism is
to be found in the present rush of experts to "solve the
problem of world hunger"—which is rarely defined except
as a "world problem" known, in industrial heroic jargon,
as "the world food problematique." As is characteristic of

industrial heroism, the professed intention here is entirely salutary: nobody should starve. The trouble is that "world hunger" is not a problem that can be solved by a "world solution." Except in a very limited sense, it is not an industrial problem, and industrial attempts to solve it—such as the "Green Revolution" and "Food for Peace"—have often had grotesque and destructive results. "The problem of world hunger" cannot be solved until it is understood and dealt with by local people as a multitude of local problems of ecology, agriculture, and culture.

The most necessary thing in agriculture, for instance, is not to invent new technologies or methods, not to achieve "breakthroughs," but to determine what tools and methods are appropriate to specific people, places, and needs, and to apply them correctly. Application (which the heroic approach ignores) is the crux, because no two farms or farmers are alike; no two fields are alike. Just the changing shape or topography of the land makes for differences of the most formidable kind. Abstractions never cross these boundaries without either ceasing to be abstractions or doing damage. And prefabricated industrial methods and technologies *are* abstractions. The bigger and more expensive, the more heroic, they are, the harder they are to apply considerately and conservingly.

Application is the most important work, but also the most modest, complex, difficult, and long—and so it goes against the grain of industrial heroism. It destroys forever the notions that the world can be thought of (by humans) as a whole and that humans can "save" it as a whole— notions we can well do without, for they prevent us from understanding our problems and from growing up.

To use knowledge and tools in a particular place with good long-term results is not heroic. It is not a grand action visible for a long distance or a long time. It is a small action, but more complex and difficult, more skillful and responsible, more whole and enduring, than most grand actions. It comes of a willingness to devote oneself to work that perhaps only the eye of Heaven will see in its full intricacy and excellence. Perhaps the real work, like real prayer and real charity, must be done in secret.

The great study of stewardship, then, is "to know / That which before us lies in daily life" and to be practiced and prepared "in things that most concern." The angel is talking about good work, which is to talk about skill. In the loss of skill we lose stewardship; in losing stewardship we lose fellowship; we become outcasts from the great neighborhood of Creation. It is possible—as our experience in *this* good land shows—to exile ourselves from Creation, and to ally ourselves with the principle of destruction—which is, ultimately, the principle of nonentity. It is to be willing *in general* for beings to not-be. And once we have allied ourselves with that principle, we are foolish to think that we can control the results. The "regulation" of abominations is a modern governmental exercise that never succeeds. If we are willing to pollute the air—to harm the elegant creature known as the atmosphere—by that token we are willing to harm all creatures that breathe, ourselves and our children among them. There is no begging off or "trading off." You cannot affirm the power plant and condemn the smokestack, or affirm the smoke and condemn the cough.

That is not to suggest that we can live harmlessly, or strictly at our own expense; we depend upon other creatures and survive by their deaths. To live, we must daily break the body and shed the blood of Creation. When we do this knowingly, lovingly, skillfully, reverently, it is a sacrament. When we do it ignorantly, greedily, clumsily, destructively, it is a desecration. In such desecration we condemn ourselves to spiritual and moral loneliness, and others to want.

Design by David Bullen
Typeset in Mergenthaler Sabon
by Robert Sibley
Printed by Maple-Vail
on acid-free paper